THE MECHANIZATION OF WORK

A SCIENTIFIC AMERICAN *Book*

THE MECHANIZATION OF WORK

W. H. FREEMAN AND COMPANY
San Francisco

The Cover

The painting on the cover symbolizes the theme of this book: the introduction of new technologies in the workplace and the social and economic consequences of technological change. The keyboard shown is that of the CGC 7900 Color Graphics Computer, made by Chromatics, Inc., of Tucker, Ga. The lower part of the keyboard is much like that of an ordinary typewriter, but an additional bank of keys controls functions that are useful in creating graphic images with the aid of the computer. For example, the operator can press the key labeled "Create" and enter a series of commands for the construction of a geometric figure; pressing the "Redraw" key causes the specified figure to appear on the screen of a cathode-ray tube. The unlabeled, colored keys assign colors to areas of a drawing; the functions of the keys in the top row can be defined by the operator. At one time the typewriter was almost the only device that had an alphabetic keyboard, and it was used mainly by clerical workers. Now the keyboard and its associated display have a new importance as the primary means of communication between people and computers. Analysts at Arthur D. Little, Inc., predict that by 1990 between 40 and 50 percent of all American workers will make use of such electronic devices.

Cover painting by Marvin Mattelson.

Library of Congress Cataloging in Publication Data

Main entry under title:

The Mechanization of work.

"The eight chapters . . . originally appeared as articles in the September 1982 issue of Scientific American"—T.p. verso.
"A Scientific American book."
Includes bibliographical references and index.
Contents: The mechanization of work / by Eli Ginzberg—The mechanization of agriculture / by Wayne D. Rasmussen — The mechanization of mining by Robert L. Marovelli and John M. Karhnak — [etc.]
1. Machinery in industry—Addresses, essays, lectures.
2. Mechanization—Social aspects—Addresses, essays, lectures. I. Scientific American.
HD6331.M378 1982 338'.06 82-17509
ISBN 0-7167-1438-8
ISBN 0-7167-1439-6 (pbk.)

The eight chapters in this book originally appeared as articles in the September 1982 issue of *Scientific American*.

Printed in the United States of America

1234567890 DO 0898765432

CONTENTS

FOREWORD

"Human labor from time immemorial," says Wassily Leontief, in his contribution to this book, "has played the role of principal factor of production." To reduce the input of labor—to ease man's labor—has supplied the economic drive to more than two centuries of industrial revolution.

First the steam engine, then the steam turbine and the internal combustion engine displaced human muscle from the tasks of production. Starting from 1820, when 70 percent of the U.S. population lived on the farm, employment in agriculture has declined to less than 4 percent of the labor force. As late as 1900, 70 percent of the population were still employed in the production of goods, 22 percent in the new manufacturing industries. Today less than 30 percent of the labor force are employed in production; more than 70 percent are in the new service occupations.

The industrial revolution has recently entered a new phase: self-regulating machines and the computer and communication technologies are displacing the human nervous system not only from its role in production processes but from service occupations as well. The change of phase is sensed, with some apprehension, in the popular celebration of "automation" and of arrival of the "robots" (from Japan, of course). There are reasons to believe, Leontief concludes, that human labor is losing and will not retain in the future its economic significance as the principal factor of production.

The necessarily momentous impact of this development upon the institutions and values of U.S. society is already felt. Shortening of the work week has halted close to an irreducible number of hours. Transfer payments (payments not for work) make up 15 percent of total personal income in this country. Talk about employment and unemployment is motivated more by concern for the distribution of goods than for production.

In the six middle chapters of this book authors actively engaged in the ongoing revolution consider the ways in which technology has reduced and continues to reduce the input of human labor in the production of goods and in the services that sustain industrial civilization. Thus, it is not only the tractor and the combine but genetics and biochemistry that make it possible for one person on the farm in the U.S. to feed 78 fellow citizens off the farm and produce a huge surplus on top of that for storage or export. Mechanization of the classical, ponderous kind has left less than 1 percent of the U.S. labor force in the mines; the preferred strategy for bringing minerals to the surface is to remove the overburden instead of tunneling through it. In manufacturing, production-line employment is principally in the tending and operating of machines; from such control functions the computer, more than the robot, is displacing human nervous systems. The value added by manufacturing from the work of white-collar employees now exceeds that from the blue-collar workers on the factory floor. The supermarket puts its self-serving customers to work, and the laser beam reading the Universal Product Code bar-codes at the checkout counter will soon make its entries directly in their checking accounts. In the office, the keyboard and the cathode-ray display tube at the "work station" tied to the company data bank upgrade the clerk to account manager. The work that remains to be done by people, from the farm to the office, is work that is increasingly worthy of their human capacities.

How an industrial economy can secure to its members the advantages of labor-displacing technology is the concern especially of the first and the last chapters of this book. The change in the kind of work people do has been accompanied by change in the character of the labor force. More than half have graduated from high school; more than 20 percent from college. More than 40 percent are female. The ranks of the unemployed, whose numbers in the U.S. have climbed to nearly 10 percent in 1982, comprise a predictably high concentration of persons not qualified for the new kind of jobs. Unemployment runs at more than 40 percent among minority youths who have dropped out of inadequate urban educational systems. Of the women in the labor force an alarmingly high percentage, employed and unemployed, are heads of single-parent households whose children are denied their promised equal opportunity. Such trends, says Eli Ginzberg, "can be disregarded only by a society that is indifferent to human deprivation and unconcerned about its own future."

If the economic problem was the wresting of man's living from the resources of the earth, it has been solved. The new economic problem is the distribution of goods and services; equity may contribute as much to its solution as economic growth.

The chapters of this book first appeared in the September 1982 issue of SCIENTIFIC AMERICAN, the thirty-third of the annual single-topic issues published by the magazine. It has been made available in book form by the expeditious work of W. H. Freeman and Company, the book-publishing affiliate of SCIENTIFIC AMERICAN.

THE EDITORS*

September 1982

1

THE MECHANIZATION OF WORK

The Mechanization of Work

by ELI GINZBERG

Introducing a volume on the continuing Industrial Revolution two centuries after it began. In the U.S. it has now displaced two-thirds of the labor force from the production of goods

The easing of human labor by technology, a process that began in prehistory, is entering a new stage. The acceleration in the pace of technological innovation inaugurated by the Industrial Revolution has until recently resulted mainly in the displacement of human muscle power from the tasks of production. The current revolution in computer technology is causing an equally momentous social change: the expansion of information gathering and information processing as computers extend the reach of the human brain. This issue of *Scientific American* is devoted to the latest stage of the historic process that has led from the most elementary force-transmitting machines to the most advanced information-handling ones.

The transformation of the U.S. labor force in the country's brief history tracks the progressive mechanization of work that attended the evolution of the agrarian republic into an industrial world power. In 1820 more than 70 percent of the labor force worked on the farm. By 1900 fewer than 40 percent were engaged in agriculture. Half a century ago, when the capitalist societies were sliding into the Great Depression, more than half of the U.S. labor force had shifted from the production of goods to the provision of services. It was then, as large-scale unemployment destabilized those societies, that national policy began to look at employment as much from concern to ensure the consumption of goods as from concern to secure their production.

Today employment in the services in the U.S. is approaching the same 70 percent that were bound to the soil a century and a half ago. Only 32 percent of the labor force are still engaged in the production of goods (mostly in manufacturing), and a mere 3 percent are employed in agriculture.

Although this transformation has been brought about largely by mechanization, it has been accompanied by social trends so pervasive that they must be included among the causes of the transformation as well as among its effects. For example, although women had begun to enter the labor force from the beginning of the Industrial Revolution, by 1980 they had come to make up 43 percent of it [see "The Mechanization of Women's Work," by Joan Wallach Scott, page 87]. The age of entry into the labor force has risen, reflecting the desire of Americans for more education and the higher level of training required by jobs in the increasingly sophisticated economy as well as the release of human labor from the tasks of production. In 1940 the median number of years of school completed by the younger members of the population was 10.3; in 1980 it was 12.9.

A disquieting feature of these dynamic internal shifts in the labor force has been the persistence of high levels of unemployment among its less educated members. Such unemployment raises the question of how any society can function effectively over the long run without bringing all its adult members into its economic life, able not only to work but also to buy [see "The Distribution of Work and Income," by Wassily W. Leontief, page 99].

The five articles that follow take up the technologies of mechanization in five areas: agriculture, mining, design and manufacturing, commerce and office work. This introductory article will of necessity deal with a limited number of themes: how the mechanization of work has been treated by economists, what its effect has been on the U.S. economy over the past few decades and what its future effect is likely to be. Particular attention will be paid to the impact of mechanization on the shifting structure and character of the labor force and on the evolution of the work environment.

Adam Smith, in *An Inquiry into the Nature and Causes of the Wealth of Nations,* published in 1776, pointed to a basic dilemma: efficiency in the generation of wealth is enhanced by the division of labor, and yet specialization that involves nothing more than routine, repetitive tasks diminishes the worker by depriving him of intellectual challenge and decision-making responsibility. Smith, preoccupied with issues of moral philosophy, expressed his concern that many workers, in a desperate effort to improve their economic circumstances, would drive themselves so hard that it would affect their health and even shorten their lives. Smith's book was written before the commercial success of James Watt's steam engine, and so Smith never had to confront the full force of modern industrialization. He nonetheless appreciated the close links between the work men do and the quality of their lives.

David Ricardo, who began his study

WORKERS AT A STEEL MILL in Pittsburgh were recorded by the noted documentary photographer and social reformer Lewis W. Hine in about 1910. Hine's revealing images of the adverse conditions of industrial labor in the early decades of this century were instrumental in the enactment of laws governing occupational safety and the employment of children. Much of the work being done by human muscle in this scene is now done by machine. The photograph, which is from the archives of the National Child Labor Committee, is now part of the Edward L. Bafford Photography Collection of the University of Maryland Baltimore County.

of political economy after reading *The Wealth of Nations* in 1799, went on to establish the classical, or free-market, school of economics. In spite of his almost exclusive emphasis on the competitive marketplace, he cautioned that increased reliance on mechanization might not turn out to be an unqualified blessing. He could see that under certain conditions workers displaced by machines might not be able to get new jobs. What was good for the employer, he concluded, might be bad for the worker.

Karl Marx devoted some of the most telling chapters in *Das Kapital* to describing the adverse effects of mechanization on the minds and bodies of working men, women and children in mid-19th-century Britain. (Because women and children received lower wages they were then replacing men in many branches of industry, from coal mines to textile mills.) According to Marx, the combination of machines,

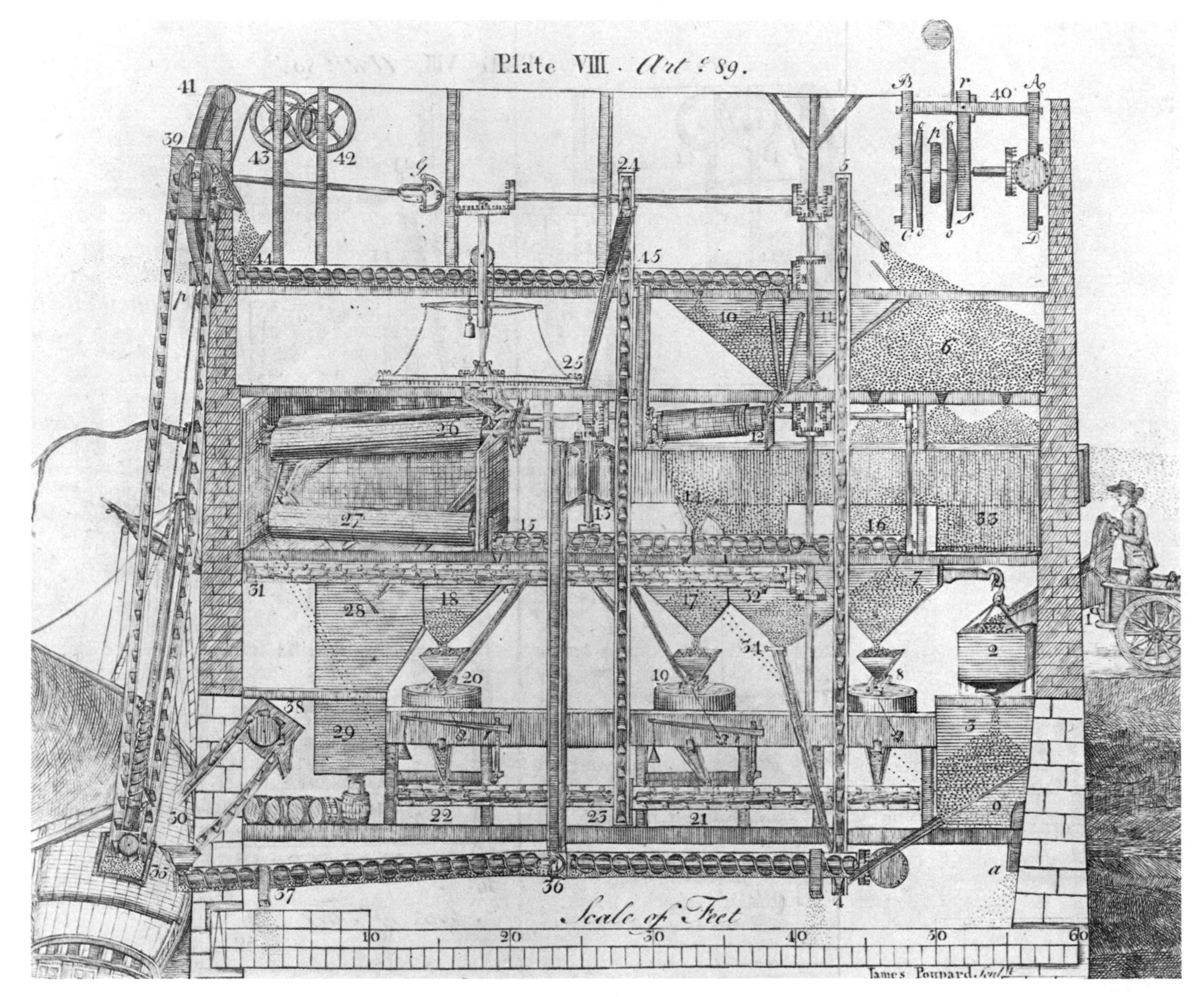

MECHANICAL FLOUR MILL patented by Oliver Evans of Philadelphia in 1790 has been described as the world's first automatic factory and the forerunner of the modern continuous production line. This schematic diagram is from *The Young Mill-Wright and Miller's Guide,* published by Evans in 1795. The mill could be supplied with grain from either a boat or a wagon. In the latter case the wagoner dumped the grain into a spout (*1*), from which it flowed into a scale (*2*) for weighing before falling into a small garner, or granary (*3*). The grain was then led to a vertical bucket conveyor (*4, 5*), which raised it to the top floor. There a crane spout could deposit it in the main storage garner (*6*), from which it could be directed into a hanging garner (*7*) that in turn fed a millstone (*8*) for rubbing, or shelling, the grain before it was ground. The rubbed grain ran by a special channel (*broken lines*) back to the first garner, where the chaff was blown through a screen into an adjacent room (*9*). The grain was again elevated to the top floor, and the crane spout was turned this time over a pair of screen hoppers (*10, 11*), which fed a rolling screen (*12*). From there the grain descended through a current of wind made by a fan (*13*). The clean, heavy grain fell through a funnel (*14*) into a horizontal screw conveyor (*15, 16*), which distributed it uniformly to the three hanging garners (*7, 17, 18*), maintaining a constant flow of grain to the millstones (*8, 19, 20*). The ground meal was moved by another screw conveyor (*21, 22*) to a second bucket conveyor (*23, 24*), which emptied it into a rotary structure called the hopper boy (*25*); this device in turn spread the meal to cool it, sweeping it gradually through holes in the floor into a room called the bolting chest, where it was sifted by a set of rotating cloth sleeves called bolting reels (*26, 27*). The superfine flour collected in a packing chest (*28*) and was led out through a spout (*29*) to fill the barrels, which could then be loaded on the boat (*30*). The coarsely ground material was removed by another screw conveyor (*31*) to a garner (*32*), which also collected the light grain blown by the fan; the chaff was driven farther by the wind and fell into a separate chaff room (*33*). The coarse material was recycled by passing it through a gate (*34*) to the bottom of the first elevator. The grain supplied by boat could be unloaded from the hold by several methods: by an articulated screw conveyor (*35, 36, 37*), by a short bucket conveyor with a fixed upper pulley (*38*) or by a long external elevator (*39*) leading to the top floor. The pulley of the external elevator was designed to rise and fall in a pair of curved slots (*41*); the mechanism for hoisting the elevator clear of the boat (*42, 43*) is shown in another view (*40*). A screw conveyor on the top floor (*44, 45*) moved the grain into the mill. Evans finished the first model of his mill in 1783, and two years later a full-scale operating version was built at Red Clay Creek near Wilmington, Del. He promoted his invention vigorously, maintaining that his improved milling machinery "lessens the expense of attendance by at least one half."

private property and competition would soon result in the self-destruction of the capitalist system. The end would come, he said, when newer and more powerful machines would drive such a large proportion of the labor force out of work that producers would no longer have enough consumers to buy the goods their machines were turning out. With the advantage of hindsight one can now see that Marx was better as a critic than as a prophet. He correctly perceived that the Industrial Revolution was harming millions of working people, but he did not allow for the substantial gains in well-being they and the generations of workers after them would enjoy because of the increased productivity resulting from mechanization.

Thorstein Veblen made technology the basis for his own penetrating analysis of modern capitalism, from his first major work, *The Theory of the Leisure Class,* published in 1899, to his last, *Absentee Ownership and Business Enterprise in Recent Times,* published in 1923. Veblen consistently maintained that the way work is organized to suit the requirements of machines determines how men think, act and dream.

In general, however, most economists—free-market, Marxist or otherwise—have failed to give technology its due. The classical theorists and their successors have built their systems and their reputations by explicating with ever greater subtlety how demand, supply and price interact in competitive markets to establish or reestablish equilibrium. To pursue this static line of inquiry they have had to ignore the influence of such dynamic factors as changes in demography, technology and taste. Moreover, because they have a limited view of efficiency they search for the margin where it pays an employer to install machines to replace workers but seldom look into such factors as the quality of the workplace and the home, both of which have come increasingly under the influence of machines.

The shortcomings in the economists' approach to the mechanization of work can help to explain many of the errors in perception and action that have characterized the U.S. economy in the period since World War II. A better understanding of the complex relations between mechanization and the economic process can be gained by reviewing some of the more important of these misperceptions and the inadequate policies they have engendered.

In 1947 the U.S. instituted the Marshall Plan. If the countries of Western Europe—both the victors and the vanquished—could agree to work together, the U.S. promised to provide them with the capital needed to speed the rebuilding of their devastated economies. Within a few years the economies of Western Europe had turned around and were growing rapidly.

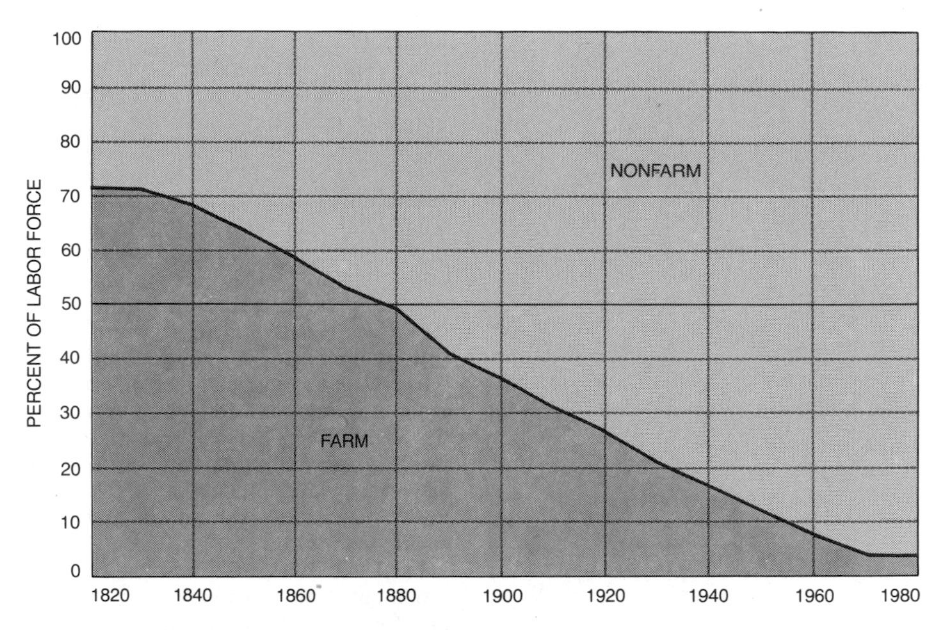

HISTORIC DECLINE in the fraction of the U.S. labor force employed in agriculture reflects the high degree of mechanization achieved on the farm in the past century and a half. In recent years agriculture in the U.S. has actually become more mechanized than manufacturing.

The success of the Marshall Plan had much to do with the inauguration of smaller-scale programs of economic assistance designed to accelerate the industrialization of the less developed countries. They too became the beneficiaries of American capital exports. Here, however, the record of accomplishment turned out to be much less impressive. Little of the so-called economic assistance went to economic development. Instead American capital exports often went in the form of arms and American dollars added to the personal wealth of those in power. Only in retrospect has it been possible to understand the reasons for the difference in outcomes. In Europe the war had destroyed factories, power plants, railroads and other facilities, but the knowledge required to run an industrial economy had remained intact. This knowledge, accumulated over a century or more, was drawn on to make good use of the new machines as soon as they were installed. In most of the Third World there was no such pool of experience, and as a result many of the imported machines were installed only after considerable delay; frequently they were operated far below capacity, and they were poorly maintained.

A second example of failure to bring

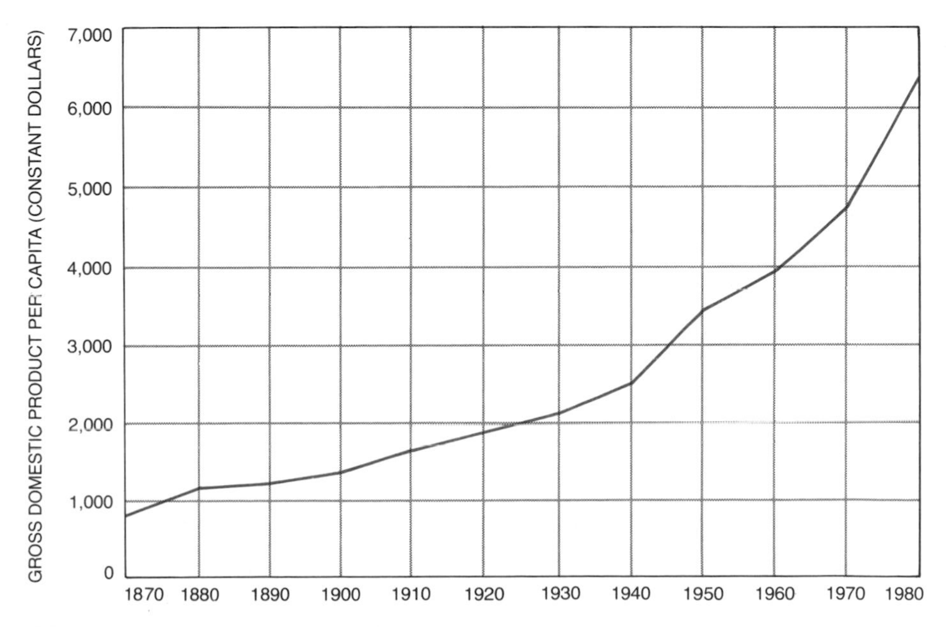

GROSS DOMESTIC PRODUCT of the U.S. has continued to rise at an approximately constant rate, when measured on a per capita basis. Nevertheless, there is considerable concern about the recent sharp decline in productivity, measured as a function of units of labor input.

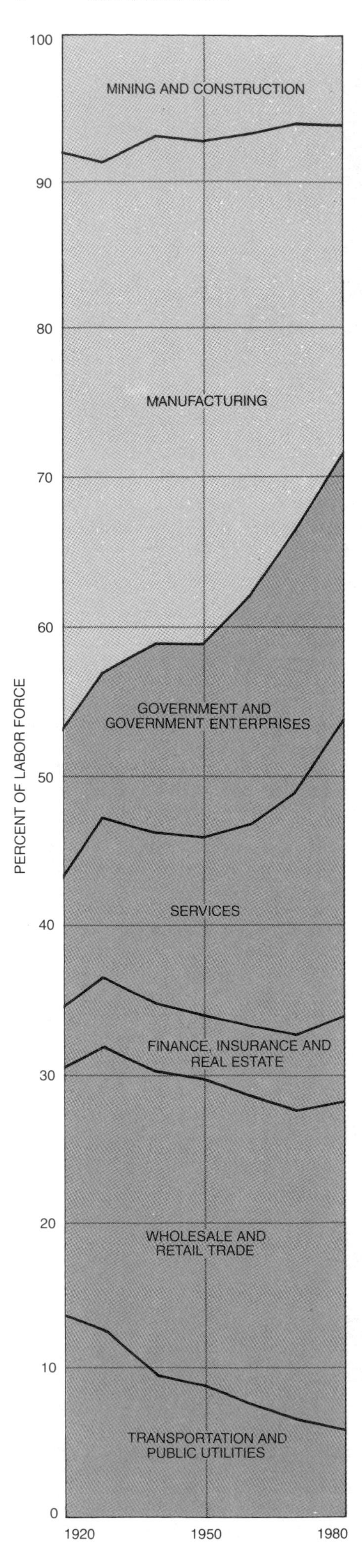

mechanization into the center of economic policy is provided by the U.S. automobile industry. Until its recent troubles that industry was looked on as the bellwether of the American economy, proof that the U.S. was the technological leader among the developed nations. Year after year the industry's sales and profits were large, and although working conditions in the assembly plants were often unpleasant and arduous, the work force was well paid and received excellent benefits. The misperception of what was happening in Detroit resulted from a widespread failure to recognize that the industry's continuing high profitability rested primarily on styling, advertising and marketing, not on advances in engineering and in manufacturing technology.

In 1962 Congress, convinced that mechanization was resulting in the disemployment of many skilled workers who would never be reabsorbed into the labor force unless they could be helped to acquire new skills, passed the Manpower Development and Training Act. That act, together with its successor legislation, the Comprehensive Employment and Training Act (CETA), passed in 1973, led to the expenditure of more than $80 billion up to the beginning of the Reagan Administration, mostly to help the poor and the near-poor. It is doubtful, however, that even 1 percent of the outlay was directed to the retraining and reemployment of workers who had lost their jobs through mechanization, because such workers could until recently make their own way into new jobs.

The most recent example of confusion about the mechanization of work arises from national economic policies ostensibly directed to "reindustrialization" (for example tax cuts for accelerated depreciation of plant and equipment, a measure expected to start a new boom in investment). The U.S. is urged to pursue other policies, public and private, that will putatively enable it to regain its eroding leadership in the manufacture of a wide range of industrial and consumer products, from steel to automobiles and television sets. Much is made of the superiority of Japanese management and the dangerous decline in the productivity of U.S. industry. However the issue is formulated, the core elements are the same: the leadership of the U.S. in technology has slipped, and there is a serious dysfunction in the attitudes, behavior and output of American workers.

GROWTH OF THE SERVICE SECTOR of the U.S. economy is represented in color on this graph for the period 1920–80. The subcategories indicated cover all wage and salary workers (including full- and part-time workers) employed in nonagricultural establishments. Included are workers in the nonagricultural goods-producing sectors (manufacturing, mining and construction) and those in the service sector, defined in the broadest sense to encompass all enterprises not engaged in the production of goods. (The subcategory "Services" is narrowly defined to designate those workers who provide services primarily to consumers.) Excluded (besides farm workers) are proprietors, self-employed people, domestic servants and unpaid family workers.

Actually the available statistics suggest that on a per capita basis the U.S. is close to its long-term trend in gross domestic product (G.D.P.): the output of all domestically produced goods and services. The unease centers on the recent sharp decline in productivity (measured as the ratio of total production to units of labor input). Any interpretation, however, is plagued by complications: the reported hours of work overstate the actual hours worked, exaggerating the measured declines in productivity; the U.S. economy has been shifting rapidly from goods to services, a shift that inadequately reflects the increases in output; the statistics also fail to adequately reflect changes in quality, investments in the public sector and what is happening outside the market, notably in the "underground economy" and in the household. If one were to understand and take proper account of these developments, the performance of the U.S. economy would probably be better, and possibly much better, than the current statistics suggest. Americans may well be unduly worried over a phenomenon that reveals more about the limitations of economic analysis and statistical reporting than about the economy itself.

The fact remains that mechanization has continued to play a leading role in the transformation of the U.S. economy and other developed economies in the past half century, as it did in the preceding century and a half. New and better machines have contributed to reducing the average weekly hours of work in manufacturing from 44 in 1930 to fewer than 42 today. At the same time mechanization has contributed to major gains in the rewards for work: the average pay in manufacturing has risen from $1.60 per hour then to $3 now (in constant 1967 dollars). This excludes fringe benefits, which have grown on the average to about 35 percent of base pay. Moreover, some economists have come to appreciate that the key to economic progress lies less with the accumulation of physical capital and more with the broadening and deepening of human capital, since it is human talent alone that is capable of inventing, adapting and maintaining machines.

Part of the problem is that the majority of economists, with their strong bias in favor of the competitive market, have paid inadequate attention to the contribution of the public sector to accelerating the growth of human capital. Public support has taken different forms: the

"G.I. Bill of Rights" of 1944, the expansion of public higher education, Federal financing of research and development, and the large-scale proliferation of specialized training programs created as by-products of efforts to build up the country's military strength and to develop nuclear power, aircraft, computers, spacecraft, communications and other large-scale technologies.

In the three decades between the election of President Eisenhower and the election of President Reagan both per capita disposable income and family income, expressed in constant dollars, almost doubled. Trade unions have become a prominent feature of the industrial landscape (although their membership as a fraction of the total work force has declined since 1955), and a professional, college-trained cadre of managers has taken command of most U.S. corporations. It would be surprising indeed if, mechanization aside, the foregoing changes had not left their mark on how workers behave both on the job and off it.

Other factors must also be taken into account: the repeated involvement of the country in foreign wars, the growing threat of nuclear war, rapid changes in basic values and behavior involving aspects of life from sex to religion, increasing skepticism about and challenges to authority and legitimacy. Only those economists who believe everything in life is determined by the calculus of the marketplace would attempt to explain the difficulties in which the U.S. economy finds itself in 1982 as resulting from a collapse of the work ethic. The Luddites looked on the machine as the villain; the supply-siders blame the worker.

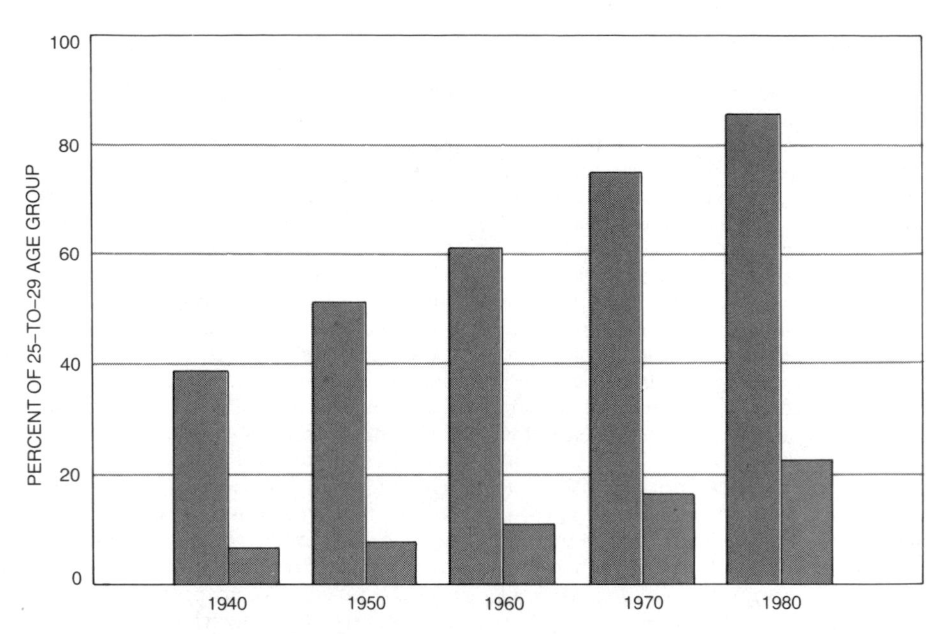

EDUCATIONAL ATTAINMENT of the U.S. population has risen markedly in the past few decades. The colored bars indicate those in the 25-to-29 age group who have finished four years of high school; the gray bars correspond to those who have been graduated from college. Between 1940 and 1980 the median number of school years completed rose from 10.3 to 12.9.

The second of the three themes I mentioned at the outset is the extent to which mechanization has helped to change the U.S. economy since World War II. Of the 41.6 million people employed in 1940 (excluding the self-employed and domestic servants) 54 percent were engaged in the production of goods: in agriculture, mining, construction and manufacturing. Mechanization had earlier made steady advances in the grain-producing states of the Middle West, but it had only a minor place in the cotton culture of the Southeast. The South, in the view of President Roosevelt, was the nation's No. 1 economic problem. It conformed to the Marxian view that surplus labor would be concentrated on the farm, living at the margin of subsistence and awaiting an opportunity to relocate to urban centers when employers needed additional workers. As late as 1940 four out of five black citizens were still living in the South, the majority of them on farms they sharecropped.

World War II was the continental divide. Many blacks went into the armed services; others moved to the North and West, where employers faced growing labor shortages; still others moved into Southern cities, many of which were being transformed by the infusion of military dollars. Other farming areas also sustained a large-scale exodus of surplus labor, setting the stage for the accelerated mechanization of agriculture. Paradoxical as it may seem, agriculture is now considerably more mechanized than manufacturing.

In the same four decades mechanization made rapid advances in bituminous coal mining as a result of two factors: the development of strip mining in the West and the decision of the United Mine Workers' Union, led by John L. Lewis, to favor higher wages over more jobs. In spite of the widespread belief that strong unions have inhibited mechanization in the construction industry, the evidence from the mechanization of excavation to the prefabrication of structures points to major advances in the application of sophisticated technologies. Although some construction unions have been strong enough to delay the introduction of new machines or to prevent the new machines from operating at full capacity, these delaying tactics have in certain instances stimulated the growth of nonunionized industry, where contractors were able to mechanize without interference.

At the height of the war boom the goods-producing sectors of the U.S. economy accounted for 69 percent of the employed labor force. In 1980 they accounted for 32 percent. The most striking shift in the goods-producing sectors was the decrease in the number, both absolute and relative, of agricultural workers. The second most prominent shift was the relative decline in manufacturing, where employment increased from 34 percent of all nonagricultural jobs in 1940 to 41 percent in 1943 but declined to 22 percent at present.

The decreasing employment in the goods-producing sectors of the economy was first matched and then exceeded by the increasing employment in the service sector. Between 1940 and 1980 employment in service occupations grew from 46 percent of total employment to 68 percent. Of all new jobs added to the economy from 1969 to 1976, 90 percent were in services.

What are the reasons for this shift? The answers differ depending on who is asked. Some economists deny that a significant shift has occurred; at most they will agree that there has been a slow, steady growth of service-sector jobs. Some acknowledge that a shift has occurred, but they ascribe it primarily to the explosive growth in health, education and related services. They expect that the growth will level off and even decline now that the birthrate is down and the Reagan Administration is pressing to reduce the level of Government outlays. Others, including our own group at Columbia University, are convinced that there has been a tilt of demand toward more consumer services and that, even more important, changes have been made in the way goods are produced, calling for a vast expansion in "producer services." Thomas M. Stanback, Jr., and his colleagues at Columbia, in their recent book *Services/The New Economy,* note that the value added of producer services alone—financial, legal, accounting, marketing, management consulting and communications—

equals the value added of all manufacturing output.

A look at the changes in the occupational structure further illuminates the causes and consequences of the shifts identified here. Somewhat simplistic comparisons can be made among white-collar workers, blue-collar workers and service-sector workers (narrowly defined as those who provide services primarily to consumers). In 1940 the proportions employed in these kinds of occupation were respectively 31 percent, 57 percent and 12 percent; in 1980 they were 54 percent, 34 percent and 12 percent. Bigger and better machines on the farm, in the mines, in the factory and at construction sites call for fewer operatives. In modern oil refineries, chemical plants and steel-fabricating mills there is a great deal of machinery but there are few workers, and many of the workers are engaged in white-collar jobs. The General Electric Company, which manufactures tens of thousands of different items from turbines to electric-light bulbs, has no more than 40 percent of its employees directly engaged in production; the rest work in what can best be classified as in-house producer services from accounting to marketing.

If one looks at the qualitative changes that are suggested by the shift from blue-collar to white-collar employment, one finds a truly impressive growth in the two groupings in the standard categories of the Bureau of Labor Statistics that have the highest status and incomes: professional, scientific and technical workers, and managerial and administrative workers. Between 1940 and 1980 the former group increased from 7.5 to 16 percent of the employed labor force, and the latter group declined from 20 to 13 percent. The last two figures conceal a major qualitative transformation, since they lump the owners and managers of small enterprises, whose numbers declined, and corporate and other high-level administrators, whose numbers rose.

Confirmation of the radical changes in the occupational structure can be found in the striking rise in the educational achievements of the younger members of the work force: those between 25 and 29 years of age. One need not hold the philistine view of many human-capital theorists that educational preparation is determined solely by the estimates people make of their career and income prospects to see that the two factors are definitely correlated. The large increase in the proportion of those in the 25-to-29 age group who have either an undergraduate degree or a higher degree is striking: from one in 16 in 1940 to almost one in four in 1980.

There is a bias among economists going back to Adam Smith that only work resulting in a physical output is productive and that services, which are by their nature ephemeral, are unproductive. Smith, reacting to the excessive number of family retainers among the rich, misled himself and his followers about the nature of services. Economists finally realized, however, that an artist who gives pleasure to thousands or a surgeon who restores the health of hundreds must be considered productive. Nevertheless, the followers of Smith have been preoccupied with refining the manufacturing model. With few exceptions the output of services has been downgraded or ignored.

This bias against service occupations was reinforced by a widespread belief that mechanization, the key to productivity and growth, has little or no role to play in the production of services. In fact, some contemporary economists have separated out the heavy, capital-intensive services—transportation, communications and electric-power utilities—and treated them as either part of or closely related to conventional manufacturing.

A further bias has been at work. Many services are anchored in the public sector rather than the private sector; the leading examples are education, health and such basic functions as police protection, fire protection and sanitation. Economic theory based on the competitive marketplace has little to contribute to an understanding of such public services. Handicapped by tradition, economists have been slow to understand the shift of modern economies toward services and in particular toward services in the public sector, toward producer services and toward mechanization in large service enterprises.

Most economists assumed that service companies would inevitably continue to be small, since service providers had to interact personally with consumers, as in the case of a restaurant, a dry-cleaning establishment, a physician or an accountant. The model of the small local consumer-service company, however, clearly does not fit the fast-food chains, the international banks with branches in 100 or more cities, the worldwide hotel chains, the national retailing chains and many other national and international service enterprises that have been able to mechanize many of their critical functions, from finance to personnel management.

As I have noted, the period since World War II has also been marked by a steady advance in the educational preparation and skill level of the work force, as exemplified by the increase in the number of white-collar workers and of professional, scientific and technical workers. The question remains of whether it is more difficult in the service sector than it is in the manufacturing sector to move from a less desirable job to a more desirable one. Stanback believes this has been the case. He points to the steelworker who began work in the yard and could move up many grades on the basis of seniority and on-the-job training. That is not the case, he observes, for the laboratory technician in a hospital or the paralegal worker in a law firm. In support of this argument, it has to be conceded that a college or professional degree is a prerequisite for competing for many of the best jobs in the service sector. On the other hand, talent appears to be as important as formal degrees in many occupations, such as advertising, design and sports. In my view the issue remains open.

These last considerations are a bridge to the third theme I mentioned at the outset: the effects of mechanization on the work environment. To the extent that any generalization is justified, one can maintain that the conventional attitude of the American worker toward machines has been different from that of

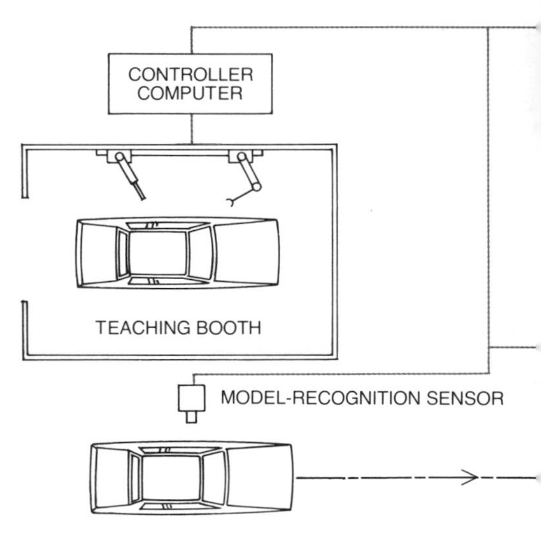

TOTAL MECHANIZATION of a new system for the spray painting of the bodies of cars and light trucks makes it possible to remove all human workers from a particularly onerous industrial task. The diagram shows the control hierarchy for the Numerically Controlled Paint System, which has been developed over the past seven years by the General Motors Corporation; the system has recently been installed at the GM assembly plant in Doraville, Ga. The present system consists of three pairs of automatic, fixed-stroke, roof

the European worker. For the most part American workers have had a positive attitude toward technological improvements, seeing them as making their work less onerous and as providing an opportunity for wage increases through increased productivity and for the enhancement of their job security through improvement of their company's competitive position.

In European countries, with their smaller markets, the job-displacement potential of the new machines has been more prominent in the thinking and action of the workers. Technological unemployment was viewed as a serious threat by the principal unions in the German Weimar Republic of the 1920's, and even the economic revival of West Germany after World War II did not dispel this fear. In the early 1960's the largest of the West German unions, the metalworkers, were host to a week-long international conference on mechanization and the involuntary unemployment it could cause. The issue is once again high on the agenda of the West German trade unions, particularly because of the disturbingly high level of unemployment in that country.

Marx railed against the dehumanization of work in which the machine set the pace, a theme that was resurrected in succeeding generations by John Ruskin, Edward Bellamy and Emma Goldman and that was developed perhaps most imaginatively in Charlie Chaplin's motion picture *Modern Times.* One need not gloss over the physical and psychological strain of working on the assembly line to point out that at the peak probably no more than one in 15 or 20 American workers earned a livelihood by such work. Robert Schrank, whose *Ten Thousand Working Days* is the most perceptive account of the diversity of working environments in the contemporary U.S. economy, makes a strong opposing case. Instead of the machine's dominating the lives of the workers, he writes, the immediate work group learns to organize its activities to enlarge its scope of freedom to do the things its members most enjoy: swap stories, fool around, play games, gamble, keep the foreman off their back and otherwise interact with one another, investing little of themselves in carrying out their assignment.

Three decades ago, in the book *Occupational Choice,* my coauthors and I distinguished three returns from work: intrinsic (direct work satisfaction), extrinsic (wages and benefits) and concomitant (interpersonal relations on the job and in the work environment). Advocates of improving the quality of work life see major opportunities to enhance the intrinsic and concomitant returns that workers are able to get from

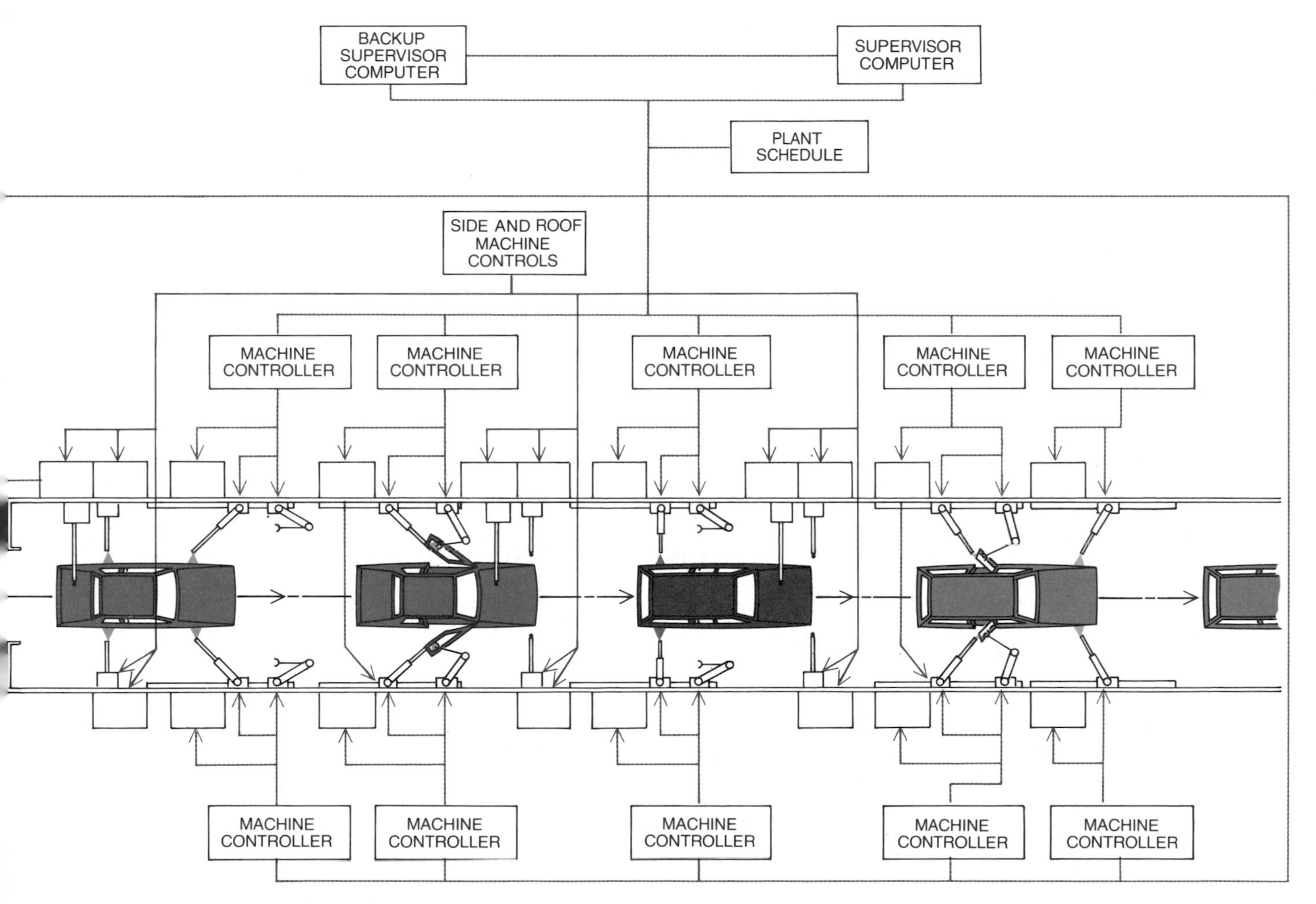

and side sprayers of a type already in wide use for such painting operations, five pairs of numerically controlled paint machines (four pairs equipped with door-opening devices) and an off-line teaching booth that houses another numerically controlled painter with its associated door opener. (The number of painting stations is expected to vary from plant to plant; one system currently being installed has 18 of the new machines.) The numerically controlled painter is a seven-axis device, hydraulically driven and servomechanically controlled. Its function is to paint all external body surfaces and various internal surfaces not covered by the roof and side sprayers. The machine's reach enables it to paint bodies of all sizes, ranging from subcompacts to full-size sedans, station wagons and pickup trucks. The painter's companion, the door opener, has two servo-controlled axes and one pneumatic axis. The supervisor computer tracks each car body through the painting booth and sends the correct path data to each machine controller at the proper time. A body-recognition system identifies each body as it enters the painting booth. The recorded information is sent to the supervisor computer and is checked against the plant schedule to determine the car's color and other options. In order to "teach" the painter a new routine a worker in the off-line teaching booth grasps a handle attached to the end of the teaching painter's arm and leads the spray guns through the appropriate paint paths, recording positions along the way and signaling "on" and "off" points. The resulting data are then stored in the system's computer.

their work. In my opinion they exaggerate. The scope for decision making by workers on the factory floor or in the large office is severely limited. An extreme division of labor results, as Smith perceived, in routine, repetitive tasks from which decision-making functions have been extracted.

Although American trade unions may have been too confrontational in their attitudes, their underlying conviction is that, beyond pressuring management to make the work environment safer, cleaner and more attractive, there is not much management can do to improve the intrinsic rewards from work. Accordingly unions have pressed and will continue to press for improvements in extrinsic rewards: job security, equity in selection for promotions, participation by the unions in discipline and discharge, better wages and fringe benefits, and more free time.

As my colleagues Ivar Berg, Marcia Freedman and Michael Freeman have documented in their book *Managers and Work Reform,* much of the agitation of the U.S. economy is a function of the expectations workers have about their jobs; there is a real danger that many are overeducated for the work to which they are assigned. Furthermore, much of the dissatisfaction of workers stems not from their limited scope to participate in decisions that affect their work but from their frustration with managers who fail to perform effectively.

Much of the preceding discussion of the workplace, worker motivation and the quality of work life has been in terms of the modern factory. Since the labor force is now overwhelmingly employed in the service sector, however, it seems desirable to call attention to a few future developments in the relation of mechanization to the work environment there.

Because of the critical importance of quality in the service sector the control of work and workers confronts management with a new and difficult challenge. Service-sector work has more dimensions and complexities than factory work, particularly considering the much higher proportion of professional, scientific and technical people employed in service industries. It is the hallmark of such personnel that their training has conditioned them to decide what work to do, how to do it and even when to do it. The members of a university faculty, although they are members of a department, a school and a larger institution, consider themselves as self-directed, autonomous individuals to whom the chairman, the dean and the president can address requests but not give orders. Increasingly this academic model is spreading to industry and government, to the research laboratories, to corporate staffs and to government agencies. There is growing tension between the traditional hierarchical structure of organizations and the implicit (and increasingly explicit) demands of professionals for greater autonomy in their work. How these demands will be reconciled with traditional modes of management remains to be seen, and the process of reconciliation may prove as difficult as it is important.

At the other end of the occupational scale it appears that the increase in the number of service-sector jobs has been correlated with the decrease in the fraction of the work force that is unionized. Many observers believe trade unions will be further weakened as the growth of the service sector continues. This may in fact happen, but several countervailing factors must be considered. Many service jobs pay low wages and provide limited benefits. More women, concentrated in low-paying service jobs, are becoming regularly attached to the work force. The computer revolution seems ready to make major inroads into the office, a development that holds a threat to the job security of many white-collar workers. The continuing erosion of the real earnings of workers by inflation makes these employees receptive to union organization. It is easy to write off the trade-union movement, particularly since it has had a conspicuous lack of success so far in restructuring itself to meet the challenges of a changing economy. Even if the unions finally succeed in making sizable gains in the service sector, they will face not only the conventional challenges of achieving higher wages and better fringe benefits for their members but also the challenge of contributing to a more stimulating workplace.

Veblen once explained the success of Germany in overtaking Britain as an industrial power in terms of the advantages of being second (or third). The latecomer did not have to carry the burden of obsolescent machinery or business practices. Many analysts in the U.S. in 1982 think Japan and the leading nations of the Third World have the same advantages Germany once had. The analogy is suggestive, but it is faulty. For some years various manufacturing activities have been moving to low-wage countries not only out of Western Europe and the U.S. but also out of Japan.

There is widespread concern about the periodic imbalances of U.S. trade in commodities with the rest of the world. In 1980 the deficit in such trade amounted to slightly more than $25 billion. That is not the whole story, however. Fees and royalties on direct U.S. investments abroad amounted to almost $6.7 billion, and net earnings on foreign investments, excluding these fees, came to $32.8 billion, resulting in a net surplus of more than $13 billion in goods and services (adjusting for the small net deficit in travel receipts). Goods and services do not lead totally independent existences, and as I have noted, services have come to play a much more important role in the production of goods. The challenge to the U.S. economy is not "reindustrialization" but rather "revitalization," in which mechanization has an important role to play with respect to both goods and services.

It is moot whether any new specific policies are required to speed revitalization beyond a recognition that the U.S. economy is moving ever more strongly into services and that the country's legislators and administrators should deal equitably between the different sectors in the creation and implementation of trade, tax and employment policies. The Reagan Administration, through the Office of the Special Representative for Trade Negotiations, has demonstrated a growing concern with international trade involving services. In the private sector a recently established consortium of major service companies is further evidence of attention and action.

A conclusion that government should not venture into the formulation of industrial policy does not imply that the state has no role to play in the strengthening of the industrial infrastructure. It is important to remember that government has played a major role in leading American industries: in agriculture, aeronautics, nuclear power, electronics, computers, communications, genetic engineering and other emerging technologies. If the present Administration has its way, the support of universities, the education and training of specialists and the underwriting of research and development will not be carried forward at an appropriate scale or with the adequate lead times. The machines that are invented, improved and put into operation throughout the economy depend on a steady accretion in the pool of knowledge and on the availability of enough technicians. If the country had to wait for the big corporations to train their own technical personnel from the ground up, it could wait a long time. Even if they wanted to do it, they could not. The ideologues may swoon over the beatitudes of the competitive market, which clearly has much to commend it, but the U.S. economy, for better or worse, is a pluralistic system in which government, nonprofit institutions and privately owned companies have complementary relations. No one of them, left to its own devices, can prosper in a technologically sophisticated world.

It would be a distortion to end this introduction to a series of articles on the mechanization of work without consideration of its problematical consequences. I shall therefore take up some of the consequences of mechanization for women and for the undereducated.

With respect to women, mechanization unquestionably paved the way for many of them to escape the confines of the home as a result of laborsaving devices, which eased the chores of housekeeping and, equally important, reduced the role of physical strength as a qualifying characteristic for many jobs. The positive role of mechanization in the liberation of women had little or no influence, however, on such untoward trends as the ominous rise in the number of households headed by women, the disturbingly large number of youngsters being brought up solely by their mothers and the large fraction of those families that live at or below the poverty level. These trends can be disregarded only by a society that is indifferent to human deprivation and unconcerned about its own future.

Before the introduction of sophisticated machinery as well as afterward all economies have faced difficulties in providing jobs for everyone who needs work. In spite of the good record of the U.S. economy with respect to the creation of jobs in recent decades Arthur F. Burns, the former chairman of the Board of Governors of the Federal Reserve System (and the current U.S. ambassador to West Germany), recommended in 1975, in the face of the continuing difficulties that many young people were having in finding and keeping jobs, that the Federal Government become "the employer of last resort" at wages 10 percent below the legal minimum wage. Some believe the shift of the economy toward services is currently making it more difficult for the undereducated to find a niche. An increasingly white-collar economy has no place for functional illiterates.

I have one concluding observation about the relation between mechanization and work. There is a widespread belief in the U.S. and Western Europe that young people have a smaller commitment to work and a career than their parents and grandparents had and that the source of the change lies in the collapse of the "work ethic." The question of why the work ethic collapsed is seldom raised, although sophisticated analysts suggest it is linked to economic affluence and the shift of concern from the family to the self.

I would suggest that the success of modern technology, which has put each of the superpowers in a position to destroy the other (and much of the rest of the world), presents a basic challenge, not only with respect to work but also with respect to all human values. It remains to be seen whether or not the potential of modern technology will turn out to be a blessing. Many young people are betting against such an outcome, and others are waiting before committing their modest stake.

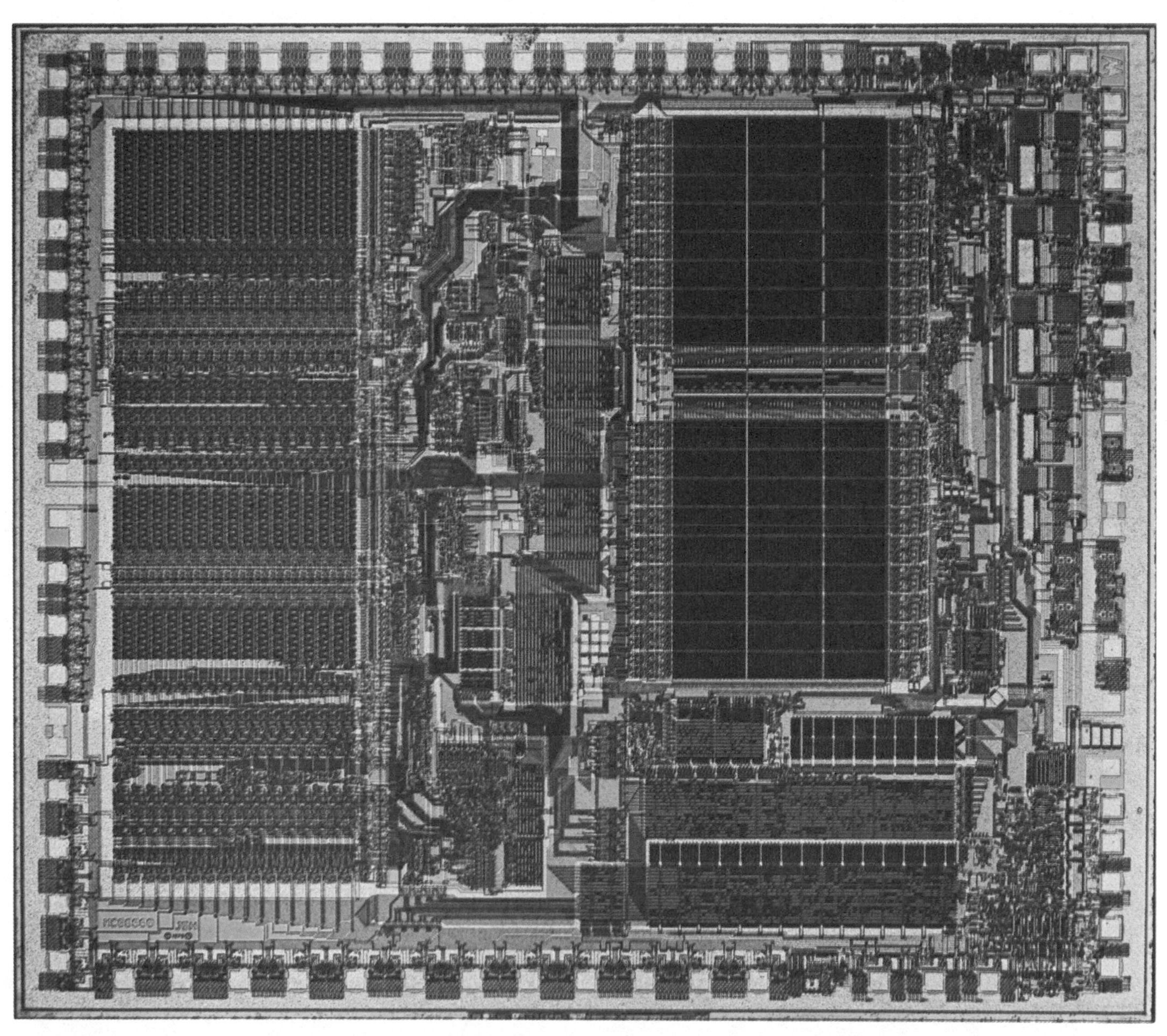

MICROPROCESSOR at the heart of the graphics computer shown in the painting on the cover crowds some 70,000 transistor sites onto a single chip of silicon measuring roughly a fourth of an inch on a side. The chip, designated the MC68000, is the first in a projected family of integrated 16-bit microprocessors developed by Motorola, Inc. The dark areas at upper right are the system's memory elements.

2

THE MECHANIZATION OF AGRICULTURE

The Mechanization of Agriculture

by WAYNE D. RASMUSSEN

In the U.S. at the beginning of the 19th century some 70 percent of the labor force worked on the farm. Today 3 percent not only feeds the population but also produces a large surplus for export

Agriculture was once the primary means of livelihood for virtually all of the human population. As recently as 1850 farmers made up 64 percent of the labor force in the U.S. Today, in contrast, only 3.1 percent of American workers are engaged in agriculture, yet they grow enough to meet the needs of the entire country, often with a large surplus for export. In 1850 the average farm worker supplied food and fiber for four people; now each farmer provides for 78 people.

Much of the enormous increase in productivity can be attributed to mechanization, broadly defined. In agriculture mechanization can be taken to include not only the introduction of devices such as plows and reapers but also the development of improved crop plants, fertilizers and pesticides, the construction of irrigation works, the growth of a transportation network for the distribution of farm produce and the extension of electric power to rural areas. These technological innovations have profoundly altered the economic and social basis of life on the farm. Indeed, the magnitude of the demographic change suggests that agriculture may be the realm where the mechanization of human work has so far had the greatest effect.

The introduction of agriculture itself, perhaps 10,000 years ago, transformed human society, and the history of farm mechanization might well be traced back to that era. In this article, however, I shall confine my attention to the past 200 years or so. Moreover, I shall consider only the mechanization of farming practices for the major crops grown in the U.S.

Land for farming has always been plentiful in the U.S., at least until now, and cheaper than labor. Hence almost any device that makes it possible to work more land with the same amount of labor has been welcomed.

At the time of the American Revolution most of the tools employed on the farm differed little from the ones that had been in use for 2,000 years. Grain was cut almost universally with a sickle, a tool that required the laborer to work in a stooped position. It was not until about the time of the Revolution that the scythe came into use; its long blade enabled the worker to cut more with one swing and its long handle enabled him to work standing up. The next improvement was the cradle, a wood frame attached to the blade of the scythe. The cradle caught the grain or hay so that it could be laid down in even rows, making it easier to gather.

With the agrarian emphasis of the 18th-century economy it is no surprise that various prominent men (many of whom had extensive agricultural holdings) were looking for new implements and more productive ways of farming. George Washington asked Arthur Young, a British advocate of agricultural change, to get improved farm implements for him. Thomas Jefferson turned his inventive mind to the improvement of farm tools and developed designs for a seed drill, a brake for separating the fibers of hemp, a threshing machine, a sidehill plow and a moldboard (the curved part of a plow blade) that would turn the soil efficiently.

The best-known advance in farm production in the years immediately after the Revolution was the invention of the cotton gin. Upland cotton, the type grown then and now in the South, has fibers that cling to the seed. Extracting the seeds had been tedious and labor-intensive work. The gin (or engine) invented by Eli Whitney in 1793 gave farmers a practical machine for separating the lint from the seeds and brought about a dramatic change in Southern agriculture. The production of cotton rose from an estimated 10,500 bales in 1793 to nearly 4.5 million in 1861. The extensive commercial production of cotton led to the expansion of the plantation system and an intensification of the system's reliance on slave labor. In this instance, then, mechanization (at least in its earliest stages) certainly did not relieve the drudgery of the worker, nor did it lead to his disemployment; on the contrary, it perpetuated the most exploitative labor practices.

In New England the availability of low-cost cotton and of the new spinning and weaving machinery developed in Britain led to the rapid industrialization of the economy. The demands of the mill towns offered farmers in New England and elsewhere an expanding market for their products, providing a stimulus to experimentation with new implements and methods. Here the economic and social effects of mechanization in agriculture began to have a direct influence on other sectors of the economy. The increased agricultural production ensured a steady supply of food at a reasonable cost to the mill workers, thus encouraging industrial development. The increased agricultural productivity meant that young people could leave the farms (indeed, they were almost forced to) for jobs in the mill towns, thereby providing industry with relatively low-cost labor while relieving population pressure in rural areas.

In cotton farming the gin virtually abolished limits on the volume of production. In the growing of grain, on the other hand, there were many bottlenecks, which were resolved only through a series of inventions and improvements in machinery. The problems began with plowing and ended with reaping.

TREND TOWARD MECHANIZATION in agriculture is evident in the photograph on the opposite page, which shows a mobile packing plant for harvesting carrots on a large farm in California. The workers at the right pull the plants and remove the tops, putting the carrots on a conveyor that takes them to a cleaning unit, where dirt and root hairs are removed, and through a washer. The workers at the left sort the carrots, putting the smaller ones in one-pound bags for retail sale and the larger ones in big fiber bags for sale to restaurants and institutions.

Plowing drew the attention of many inventors. The first U.S. patent issued for a plow went to Charles Newbold of New Jersey in 1797. His plow was a single piece of cast iron except for the handles and the beam by which it was hitched to the draft animals. It is said that farmers would not buy the plow in the belief the iron would poison the soil and make weeds grow. In 1814 Jethro Wood patented another cast-iron plow, which he improved in 1819; the moldboard, the share (the part that cuts the furrow) and the landside (which guides the plow along the furrow) were cast separately, and the three parts were interchangeable from one plow to another. Wood's plow was widely adopted.

Neither wood nor cast-iron plows worked well in the sticky soil of the prairies, which were to become the heartland of American grain farming. The soil stuck to the plow instead of sliding by and turning over. In 1833 John Lane, a blacksmith in Lockport, Ill., began fastening strips of steel of the kind meant for saw blades over wood moldboards. His plows turned furrows in the prairie loam of Illinois, but he did not patent his idea. In 1837 John Deere, who was also a blacksmith in Illinois, began making plows out of saw steel and smooth wrought iron. The plows were highly effective in the prairie soil, and Deere, in partnership with Leonard Andrus, soon built up a substantial business.

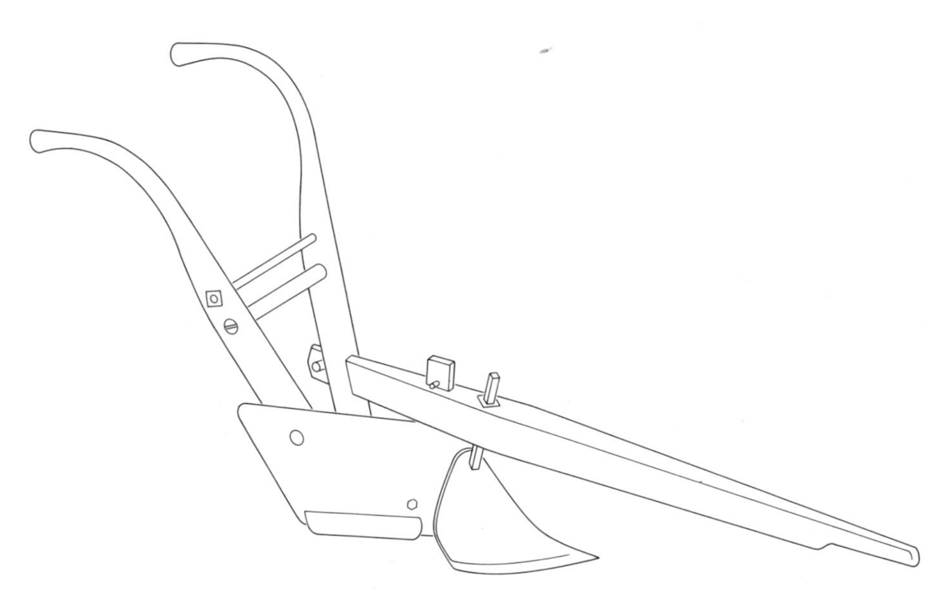

EVOLUTION OF THE PLOW in the U.S. over the past 200 years is traced in the illustrations on this page and the opposite page. This horse- or ox-drawn plow, which is typical of the kind used in the 18th century, was made out of wood except for the pointed share at the front.

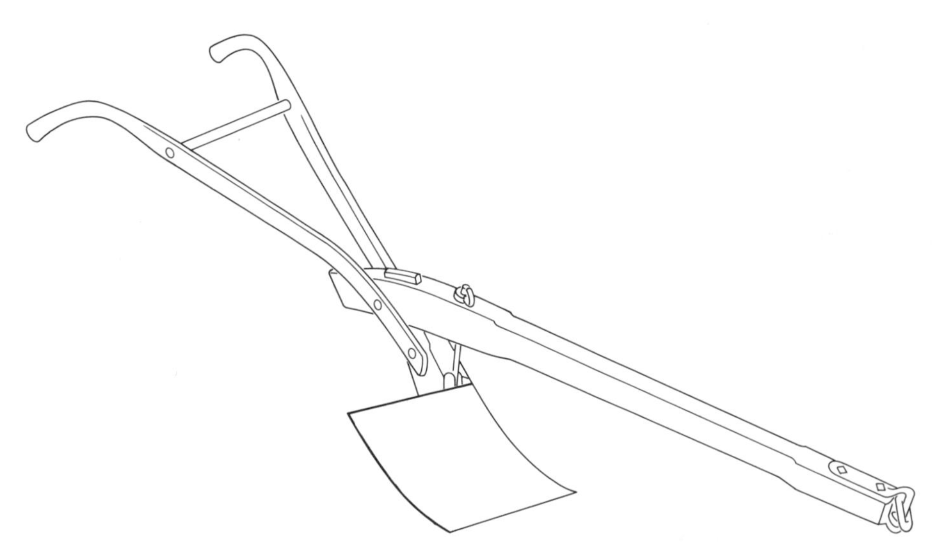

PRAIRIE PLOW was invented by John Deere in 1837 to cope with a problem that had arisen as agriculture expanded into the prairies of the Middle West: the soil stuck to the wood and cast-iron plows employed at the time instead of sliding by and turning over. Deere made his plow blade of saw steel and smooth wrought iron; it functioned well in the sticky prairie soil.

Harvesting was the crucial operation in the production of grain. Hence the mechanical reaper was probably the most important invention introduced into farming between 1830 and 1860. Obed Hussey of Maryland patented a practical horse-drawn reaper in 1833. Cyrus H. McCormick of Virginia had devised a similar machine by 1831, continuing work along lines begun by his father, and he patented it in 1834. Over the next 20 years McCormick gained a dominant position in the business, partly because he moved his company to Chicago, where he could better reach customers in the newly opened prairies and plains, whereas Hussey kept his plant in Baltimore.

By 1851 McCormick was producing 1,000 reapers a year in his Chicago plant. Over the years the harvester was improved in various ways. One notable improvement was the twine knotter, which was perfected by John F. Appleby in 1878. It enabled the machine to tie the cut grain into bundles for quicker and easier handling.

The success of the reaper and other horse-powered devices (including some developed earlier than the reaper) encouraged a trend toward machines that did not depend on human muscle power. A corn cultivator and a hay and grain rake, both drawn by horses, were available by the 1820's. In 1837 John A. Pitts and Hiram A. Pitts patented a commercially successful threshing machine. A mower that gained wide acceptance was patented by W. F. Ketchum in 1844. Other horse-powered machines marketed before the Civil War included grain drills, corn shellers, hay-baling presses and cultivators of various types.

In spite of these advances, which were publicized and advertised in the agricultural magazines of the 1840's and 1850's, many farmers hesitated to invest in the new machinery until they could be certain the investment would pay off. It was the Civil War that ultimately provided the impetus for change, and the subsequent conversion from hand power to horsepower can be designated the first American agricultural revolution. The progress of the revolution is recorded in the statistics on investment: in constant-value dollars the average annual investment per farmer in new machinery and equipment was $7 in 1850, $11 in 1860, $20 in 1870 and $26 in 1880. The labor shortage brought on by the war, together with high prices and a seemingly limitless demand, encouraged farmers to spend their savings or to go into debt to acquire the labor-saving machines. Once the farmer had made the investment he found himself committed to production on a commercial scale.

From 1820 to 1850 there had been virtually no change in the productivity

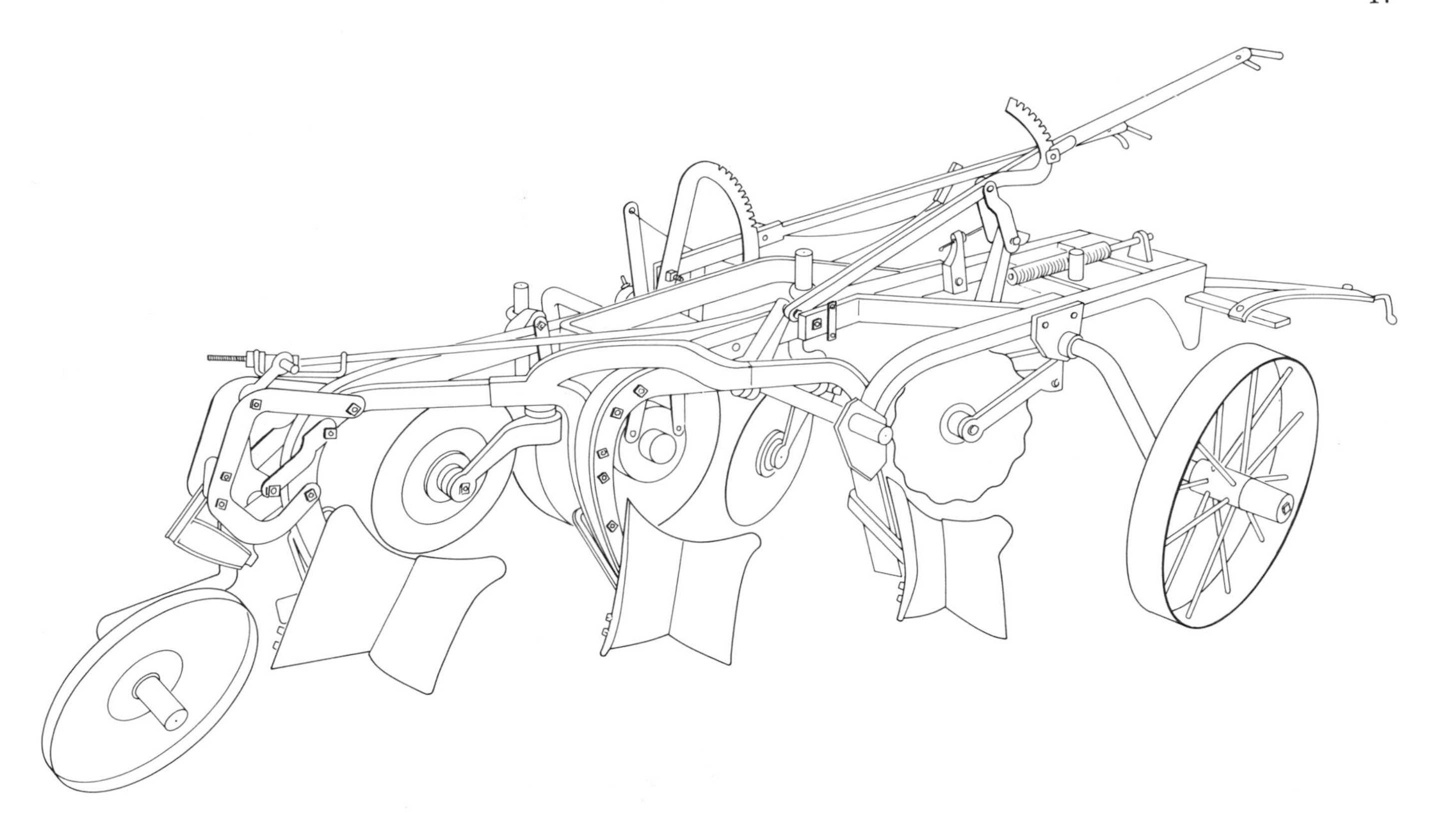

TRACTOR PLOW of the type that was common in the 1920's represents an early stage in the transition from similar horse-drawn apparatus. In the horse-drawn version the plow had a seat for the farmer. Many horse-drawn plows continued in use until after World War II.

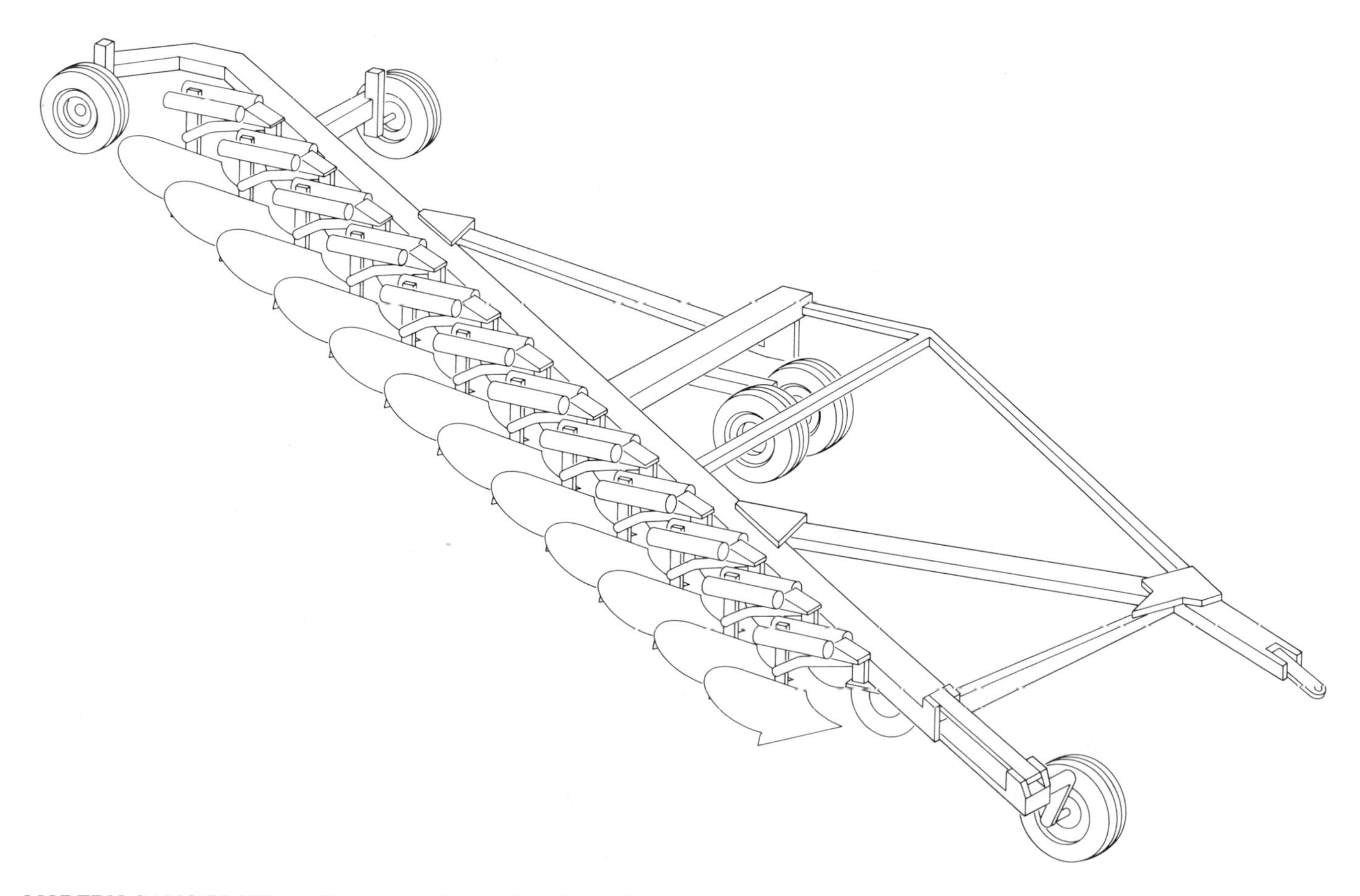

MODERN GANG PLOW has 12 blades, and each blade is capable of making a 22-inch cut. When the device is drawn by a tractor on level, open ground, it can plow 10 acres per hour or more. Deere & Company makes this plow and others with from 10 to 16 blades.

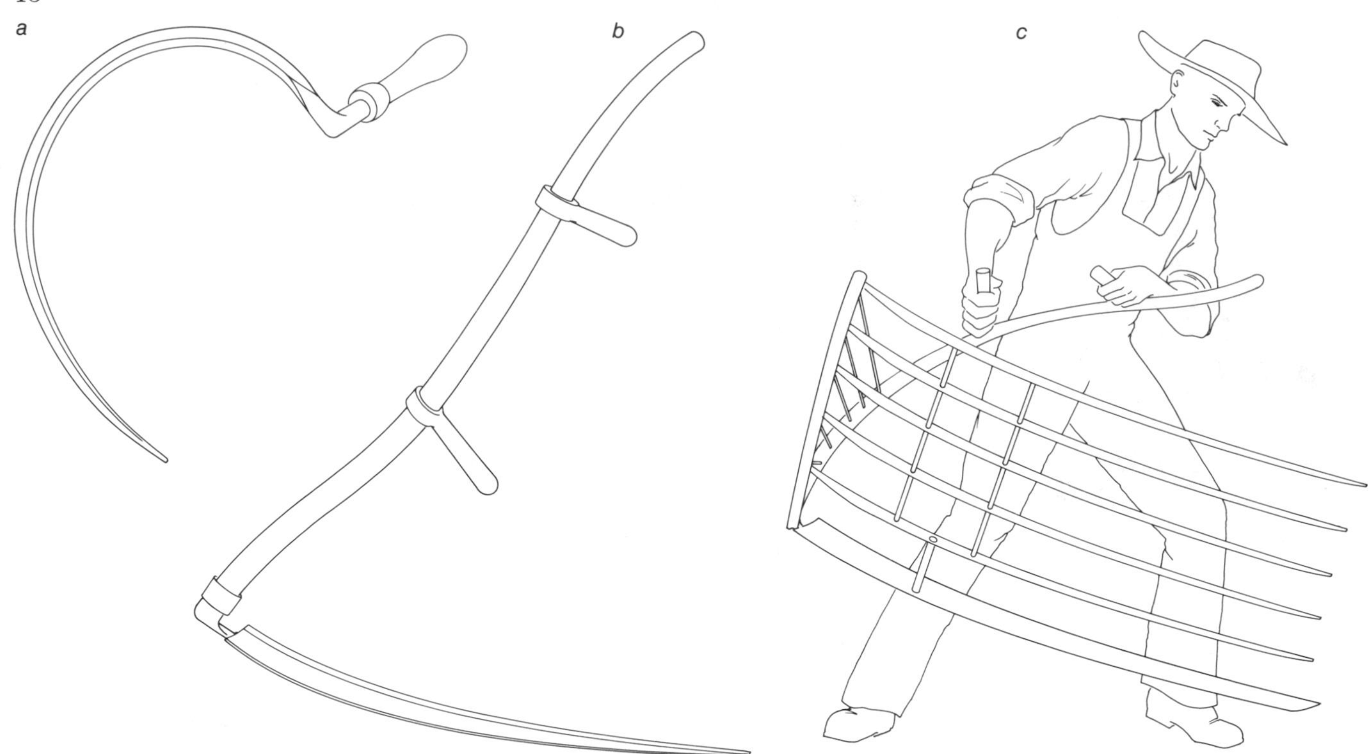

EARLY REAPING TOOLS were in use in the U.S. until well into the 19th century. The sickle (*a*) required the harvester to cut grain or grass in a stooped position; the scythe (*b*), developed at about the time of the American Revolution, enabled him to work standing up and to cut more with each stroke. The cradle (*c*) was the most advanced harvesting implement known to farmers of the early 19th century. It was a scythe with a wood frame attached. With it the harvester could catch the cut grain or hay and lay it down in even rows.

of the farm worker. In the following 30 years, however, production per man-year increased markedly, even though the average yield per acre improved little. The number of people who could be supplied with food and fiber by one farm worker rose from four in 1850 to five and a half in 1880. The number of farmers continued to increase, but at a rate lower than that of the general population. As a result farm workers as a proportion of the U.S. labor force decreased from 64 percent in 1850 to 49 percent in 1880.

The technological changes also had social repercussions. The capital needed to establish a farm increased, making it harder for laborers, tenants and young people to become operators of farms. Farmers became more dependent on bankers and merchants. Except during

THRESHING OPERATION in Kansas in 1909 was done with steam-powered machinery but still required a considerable work force. The steam tractor at the right, with a coal-carrying trailer attached to it, delivered power to the separator by means of the long

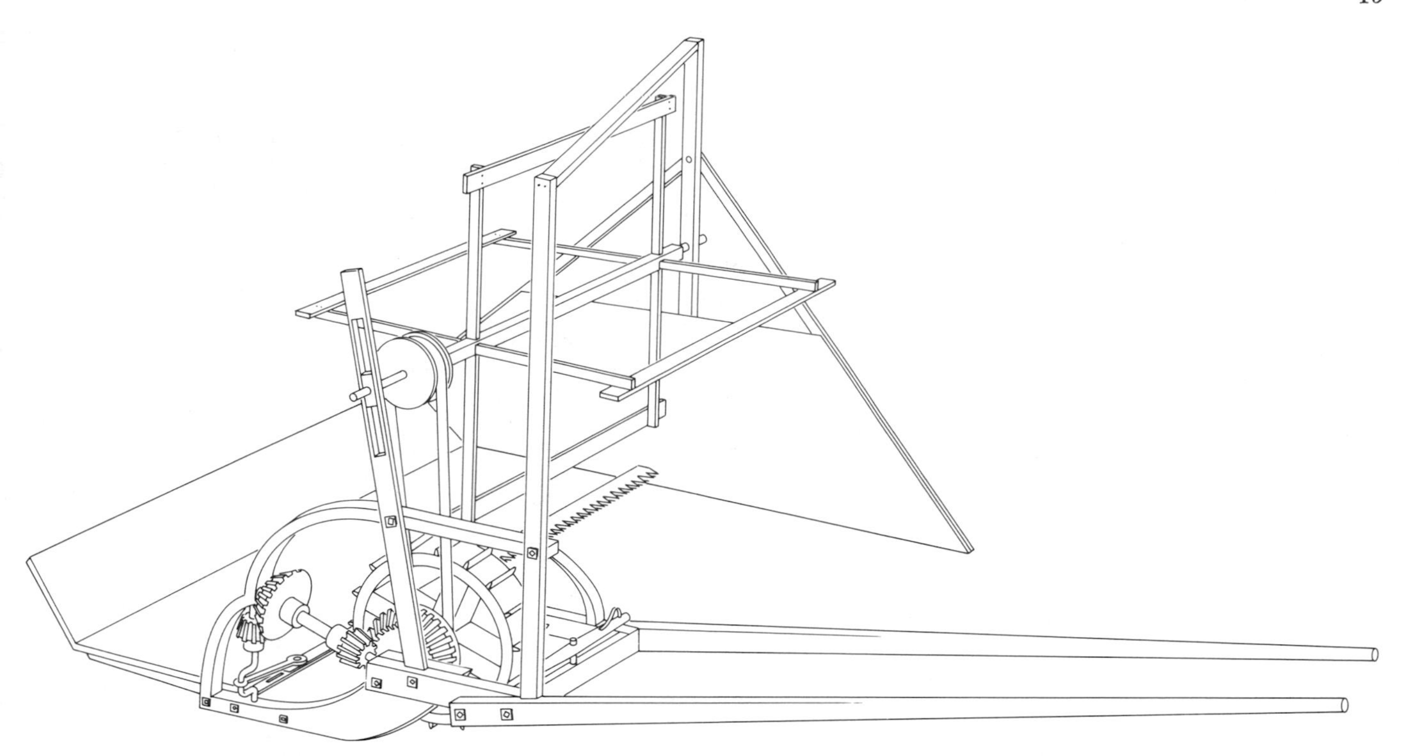

MECHANICAL REAPER was patented in 1834 by Cyrus H. McCormick. It was not the first reaping machine, but it soon became the dominant one. In this one-horse model the large ground wheel transmitted power to the reciprocating cutter blade by means of gears and a crank and to the revolving reel by means of a belt. The reel gently pushed the grain onto the platform at the rear. The grain was then raked onto the ground by hand and formed into small sheaves that were picked up and gathered into shocks to dry before threshing.

the war years the higher production of agricultural goods led to periods of surplus and low prices. Farmers were advised to reduce production, but no one farmer could influence the market. Furthermore, it was necessary to bring more land under cultivation in order to pay for the very machines that made this expansion of agriculture possible. For at least some farmers mechanization undoubtedly brought release from toil and a measure of prosperity, with the associated benefits of more leisure and better education. To some extent, however, it seems the agricultural bounty of the late 19th century was secured at the expense of the farmer.

The Homestead Act of 1862, which made Western land available free to settlers, and the building of the transcontinental railroads encouraged established

belt. The separator removed the grain from the stalks. Five horses (of the 10 or so needed in the operation) and 13 men can be seen. Most of the men worked with pitchforks, moving shocks of cut grain onto wagons in the field and then pitching the grain into the separator.

farmers and immigrants from Europe to go west, opening up the plains first to cattle and then to wheat. On the newly cultivated land machines became increasingly important. And even as the horse became fully established as the prime mover of agriculture, new sources of power were developed. Steam engines, first stationary and then self-propelled, were applied in many operations on large farms, particularly in the West.

The Red River Valley in North Dakota and Minnesota provides an impressive example of large-scale farming based mainly on steam power. When the Northern Pacific Railroad suspended its construction of new lines in the major economic depression that began in 1873, some of the officials of the company accepted land in the Red River Valley in exchange for their railroad bonds. In 1875 Oliver Dalrymple, an experienced wheat farmer in Minnesota, contracted to manage the land. He began raising wheat on tracts so large that some of the plowed furrows were six miles long. Although horses and mules supplied much of the power, Dalrymple put his emphasis on steam tractors, which he hitched to the most modern machinery available.

Dalrymple's venture provided an early demonstration of the problems that can beset a large mechanized farm operation: breakdowns of machinery, the unreliability of labor, variations in climate and low prices for farm products. The same problems affected smaller, family-operated farms, but the family farm had more opportunity to reduce or defer expenditures because the family did not have to meet a payroll for hired labor or show a profit for the owners. By the 1890's most of the large, single-crop farms had given way to smaller, diversified holdings.

In general the task for which steam engines proved to be most useful was threshing grain. The engines were too heavy and cumbersome for most other farm work. The peak in the manufacture of steam engines for agriculture came in 1913, when 10,000 of them were made. Production declined rapidly thereafter as the gasoline-powered tractor began to dominate the market.

The first practical self-propelled gasoline tractor was built in 1892 by John Froelich of Iowa. He mounted a gasoline engine made in Cincinnati on running gear fitted with a traction arrangement of his own manufacture. With this machine he made a 50-day threshing run. The Froelich tractor was the forerunner of the Deere line of tractors.

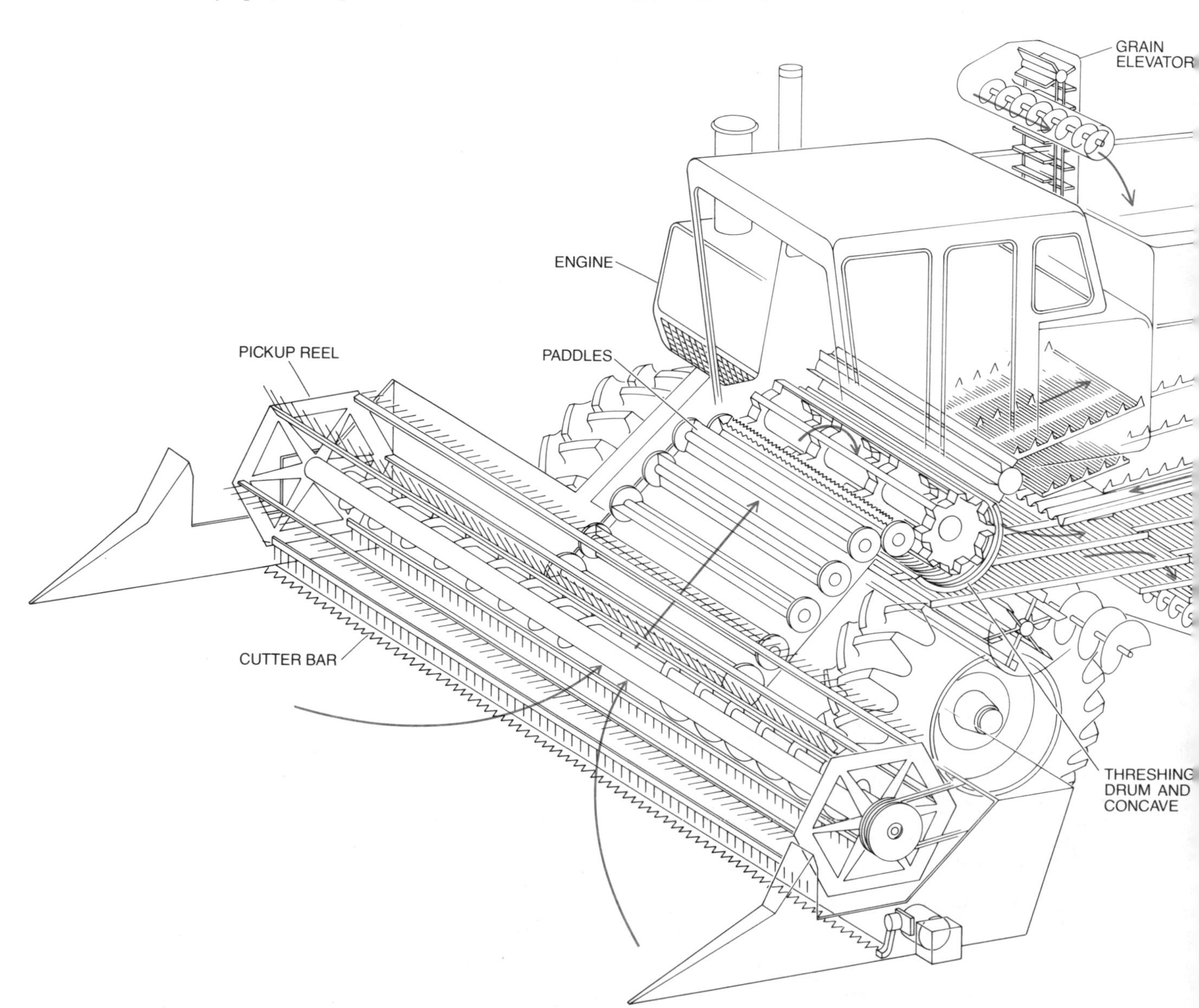

MODERN COMBINE derives its name from the fact that it combines the operations of harvesting grain and threshing it. The grain is cut by the reciprocating cutter bar and picked up by the reel. Augers steer the grain stalk-first onto a moving elevator that carries it to the threshing drum, where it is stripped and beaten. Some grain drops into the grain pan. The straw and the rest of the grain go on to the strawwalkers, which shake back and forth and cause more grain to fall into the pan. The grain elevator empties into a bin that can hold about 180 bushels of grain until the auger at the rear unloads it into a truck. The straw drops off the walkers and onto the ground; the chaff

The first business concerned exclusively with the manufacture of tractors was established in Iowa City in 1905 by C. W. Hart and C. H. Parr. They had started working on internal-combustion engines after meeting as students at the University of Wisconsin in 1893. Their first tractor came out in 1901; it was crude, but it remained in service for 20 years. The Hart-Parr Company later became part of the Oliver Corporation. Many other tractor-making concerns (not all of them successful) were formed in succeeding decades.

The adoption of the gasoline tractor spread rather slowly until World War I. Then high prices for farm products, government appeals for increased production and labor shortages in some areas encouraged farmers to make the transition. After the war and a sharp drop in farm prices in July, 1920, the pace of conversion slackened again. Through the rest of the 1920's fluctuating economic conditions in agriculture made farmers reluctant to give up horse-drawn equipment for the tractor, which would entail additional outlays of cash. Nevertheless, the number of horses and mules on farms decreased steadily and the number of tractors increased.

One machine that had come into its own by the 1920's was the combine, which cuts the grain and threshes it in a single operation. In the harvesting of wheat the combine replaced both the stationary threshing machine and the grain binder, which had mowed the wheat and bundled it into shocks. The first successful combine was a horse-powered machine built in 1836 in Michigan. By the 1880's steam engines had been introduced to power the many combines coming on the market. The gasoline engine began to replace steam for pulling the combine and operating its harvesting mechanisms in about 1912. Large combines powered by gasoline engines were widely available in the 1920's and 1930's. The development in 1935 of a one-man combine powered by a two-plow tractor (that is, a tractor with enough power to pull two plows) was another milestone.

By 1956 more than a million combines were at work on American farms, and the 1.5 million grain binders that had been in use in the decade before World War II had virtually disappeared. Also gone was the substantial crew that had made threshing and the feeding of threshers the biggest and hardest jobs of the year on most grain farms. The threshing crew itself usually consisted of three men. The separator man was responsible for the operation of the machine that separated the grain from the straw and the chaff. The engine man ran the tractor, which powered the separator by means of a long belt. The steam tractor remained in service for threshing long after the gasoline tractor had become common for other work, and so the crew also had a water man to supply the steam tractor with water. If the threshing was done from shocks in the field, two or three more men were needed to pitch bundles of grain onto horse-drawn hayracks, each of which had a driver, and often another man or two men were needed to pitch bundles into the separator. The crew for a farm of medium size would run to about a dozen workers. Those who were not in the family and not living nearby could sleep in the barn, but they all had to be fed. Cooking for them was the ultimate test of the farm housewife.

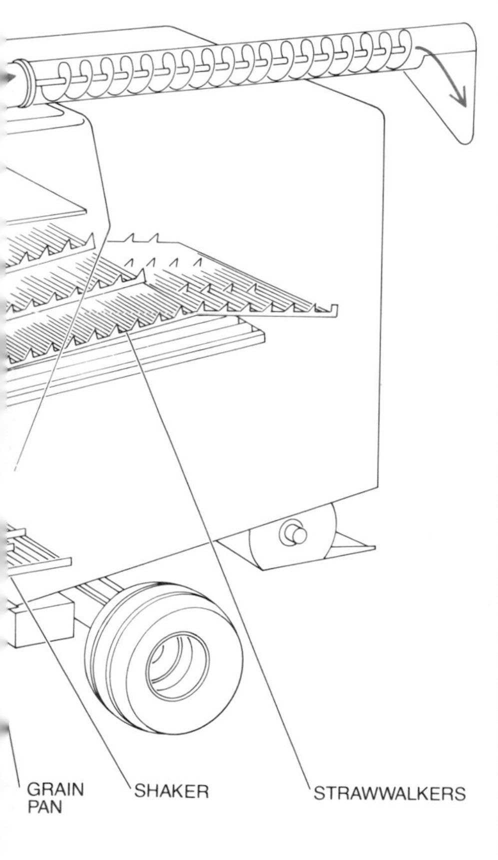

is blown out at the rear by a fan. The entire operation is powered by a diesel engine. The "concave" fits around the threshing bars. This large combine, which can harvest 12 acres per hour, is made by Massey-Ferguson, Inc.

World War II was the impetus for the virtually complete transition to mechanization. This was the second American agricultural revolution: the change from animal power to mechanical power. The war was not the only factor leading to change. In the 1930's the farm programs of the New Deal had enabled some farmers to replace worn-out machines with new models. The rural-electrification program had brought a major new source of power to many farms (and eventually to nearly all of them). With electricity farmers could run useful devices of all kinds, including not only electric lights but also milking machines, feed grinders and pumps. It took the war, however, and the accompanying shortages of farm labor, high prices for farm products and an enormous demand for those products to convince nearly all American farmers to turn to tractors and related machines.

In general American farmers have adopted a new technology on a large scale only when it has been developed and tested at a time of favorable economic conditions. The early inventions often began in the shops of farmer-mechanics, but the testing, improving and selling were done by the manufacturers of farm implements. Later on engineers at the land-grant universities and at the Department of Agriculture built the prototypes of new machines; again, however, the manufacturers did the testing, perfecting and distributing, as well as some of the inventing. The land-grant universities, the agricultural-experiment stations and the extension services have had another essential role: they have educated farmers in the advantages of the machinery and in the way farm work should be adjusted to exploit it fully.

New machines were only one aspect of the second agricultural revolution; mechanization and many other changes constituted a package of practices, or what has been called the systems approach to the improvement of agricultural productivity. The other changes included the controlled application of lime and fertilizer, soil-conservation techniques such as the planting of cover crops, irrigation where necessary, the creation of improved varieties of plants and breeds of animals, the adoption of hybrid corn, the formulation of more balanced feeds for livestock, the more effective control of insects and diseases and the use of chemical weed killers and defoliants. The effects of these practices on agricultural output were dramatic. In many respects rural life was affected just as profoundly because of the consolidation of farms and the sharp decline in the farm population.

The production of sugar beets is a good example of laborsaving through mechanization and the development of a new type of seed. Before World War II the sugar-beet farm had a few machines and specialized implements, including horse-drawn seed drills, cultivators, lifters (for pulling up the plants) and wagons, but hand operations dominated thinning in the spring and harvesting in

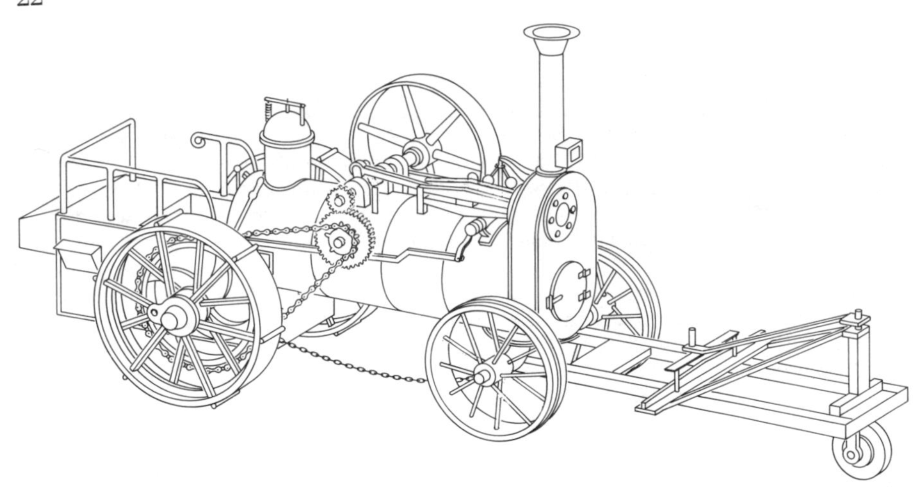

FARM LOCOMOTIVE ENGINE was one of the earliest steam tractors; it became available in about 1860. The railed area at the back is a fuel bunker. The one-wheel rig at the front is the steering apparatus; an operator sitting on the board seat turned the single wheel by means of the vertical handle in front of the seat, causing the front wheels of the tractor to turn. Another man ran the boiler. A third man was probably needed to keep the boiler supplied with water.

the fall. The work was generally done by migrant laborers. In the 1930's the California Agricultural Experiment Station at Davis and the Bureau of Agricultural Engineering of the Department of Agriculture began a cooperative research effort aimed at developing a combine for sugar-beet harvesting, that is, a machine that would lift the plants, remove the tops and load the beets in one pass down the row. Over a period of years several devices were developed to do one part of the job or another; the ideas were turned over to private manufacturers, who carried on the work. By 1958 two major types of harvester were being manufactured. In that year the entire sugar-beet crop was harvested mechanically, compared with only 7 percent in 1944.

The task of thinning out the weaker plants was tackled both mechanically and by modification of the seed. In 1941 some multigerm seeds, which ordinarily give rise to a cluster of several plants, were sheared into segments and planted; many of them produced only one plant, greatly diminishing the need for thinning. Sheared seeds were adopted by a number of growers. In 1954, however, a better solution was made available with the release to growers of the first monogerm seeds. Today, with the precision planting of monogerm seeds and the use of a mechanical thinner, the production of sugar beets in the U.S. is completely mechanized. The growers have become independent of migrant labor.

The migrant-labor problem was also back of the research leading to a mechanical tomato harvester. Much of the tomato crop in California had been picked by Mexican laborers who entered the U.S. under the terms of the Bracero program. When the program was ended in 1964, growers reported it was not possible to recruit U.S. citizens to do the work. Some labor leaders disputed this view, but the controversy was effectively ended by the successful mechanization of the harvest.

The application of machinery to tomato picking required the convergence of two lines of work, both carried on at the California Agricultural Experiment Station at Davis. One achievement was the breeding of a new variety of tomato plant by Gordie C. Hanna. With appropriate amounts of fertilizer and water the plants of this variety set an abundant crop of fruit, and all the fruit ripens at about the same time, so that an entire field can be harvested at once. Furthermore, the fruit can withstand machine handling.

The harvesting machine itself was devised by Coby Lorenzen, Jr. It cuts the plants at ground level, lifts them into a compartment and removes the fruit by shaking the vines. A belt then carries the tomatoes past a crew riding on the machine, who remove green fruit and clods of dirt. The first tomato harvesters had a capacity of from eight to 12 tons per hour. A total crew of 12 handled about the same amount of fruit that could be harvested by 60 people picking manually. In 1963 only 1.5 percent of the tomatoes grown in California for processing were harvested by machine; by 1968 the fraction was 95 percent, and now it is virtually 100 percent. Moreover, the crew on the machine has been reduced to three or four.

The mechanization of cotton harvesting has affected even more people. The first device, the cotton stripper, removed the bolls from the plant. It came into wide use in Texas in 1926 but was reasonably satisfactory only in certain areas. In 1928 John D. Rust and Mack D. Rust of Texas patented a spindle picker, which pulled the cotton fibers from the boll by wrapping them around spindles. The spindle picker gained acceptance slowly until World War II, when high demand and high prices for cotton stimulated manufacturers to build more machines. Even so, less than 10 percent of the cotton crop was harvested by machine in 1949. By 1969, however, the fraction was at least 96 percent. In the same period improvements in land preparation, planting techniques, the application of fertilizers and the control of water, weeds and pests increased the average yield of upland cotton from about 300 pounds of lint per acre to more than 500. Most of the drudgery had been eliminated from cotton farming. In 1945 about 42 man-hours were needed to produce 100 pounds of cot-

FROELICH TRACTOR, made by John Froelich of Iowa in 1892, was the first practical self-propelled gasoline tractor. It was also the first to propel itself both forward and backward.

ton in the U.S., but by 1975 the same output was being achieved in only two-thirds of a man-hour.

The sharp gain in productivity has reduced by several hundred thousand the number of workers needed on American cotton farms. One effect of the diminished need for manpower has been the virtual end of sharecropping, which had long been considered detrimental to the sharecroppers, the owners and the land. Those farmers who continue to raise cotton have benefited from a rise in the rural standard of living and from a lessening of racial discrimination. On the other hand, many of the displaced sharecroppers and laborers have found few opportunities for alternative employment.

Not long after they had invented the spindle picker the Rust brothers recognized that the machine could put "75 percent of the labor population out of employment." They were unwilling to see that happen and resorted to one plan after another to prevent it. Their ideas included adapting the machine to small farms, marketing it with restrictions on how it could be deployed, selling it only to community farming projects organized as cooperatives and setting apart some of their profits to assist displaced cotton farmers. None of the schemes proved practical, however, and all of them were swept aside when the effects of World War II brought several more companies into the business of manufacturing cotton harvesters.

Another agricultural technology that has changed greatly in the past few decades is irrigation. A person flying from east to west across the U.S. can see, beginning at the line of 100 degrees longitude or just beyond it, one circular field after another, extending from Texas to North Dakota. The circles result from one of the newest developments in irrigation: the center-pivot sprinkler, an almost totally automatic system. The sprinkler is a long pipe with nozzles at intervals along its length; the pipe is mounted on wheels and pivots around the center of the field, tracing out a circular area over the crop plants. The system is propelled by water turbines or electric motors. One center-pivot sprinkler irrigates a circular area of about 133 acres.

The center-pivot system is far removed from the ditch built by the first Mormon pioneers in Utah to divert water from City Creek to their fields. This first instance of modern irrigation agriculture was based on early irrigation systems developed by the Indians of the Southwest long before the arrival of Europeans. Mormon settlers moving from Utah into parts of Arizona, Colorado, Idaho, Nevada and New Mexico took their cooperative irrigation practices with them. The technology was soon adopted throughout the West. Streams were dammed and the stored water was carried by ditches to leveled land to increase productivity far beyond what was possible with natural rainfall.

The streams and the amount of water available were severely limited in most parts of the West. In addition irrigating with surface water moved by gravity was a task demanding many hours of heavy labor as the irrigator attempted to control the movement of the water over his fields. After World War II the use of movable aluminum pipes aided in this task. The supply of water has been augmented in many parts of the West, notably in Texas and Nebraska, by massive pumping from relatively shallow wells reaching supplies of ground water built up over hundreds of years. The water is usually distributed over the land by some type of sprinkler system, of which the center pivot has proved to be one of the most efficient. The prototype of this device was built in 1949 by Frank Zybach, a Colorado tenant wheat farmer.

The major disadvantage of the center-pivot system drawing on well water in the arid West is that the supply of under-

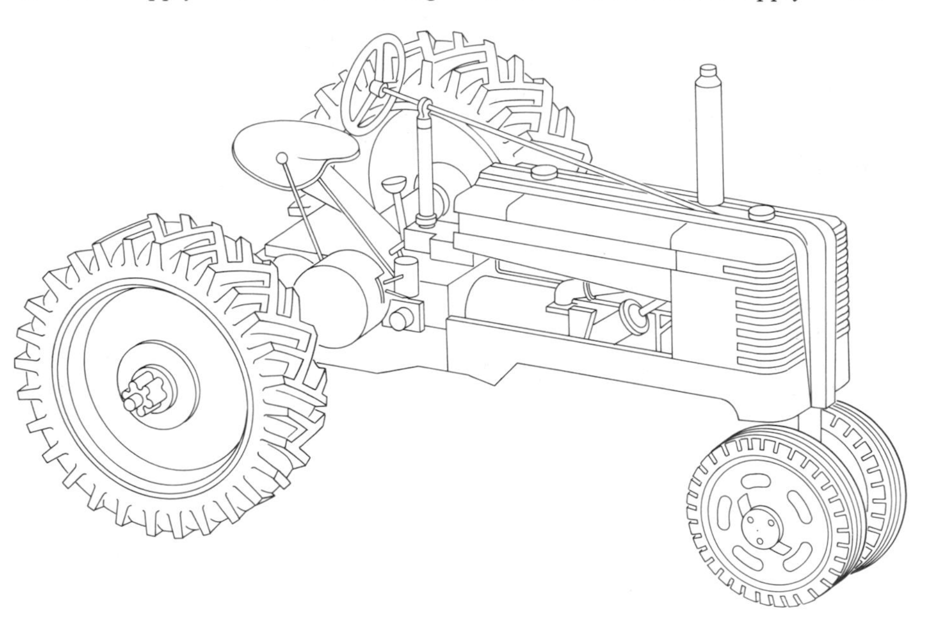

EVOLUTION OF THE TRACTOR is shown by this model of the 1930's. The front wheels were spaced to pass between two rows of plants and the rear wheels to straddle the rows.

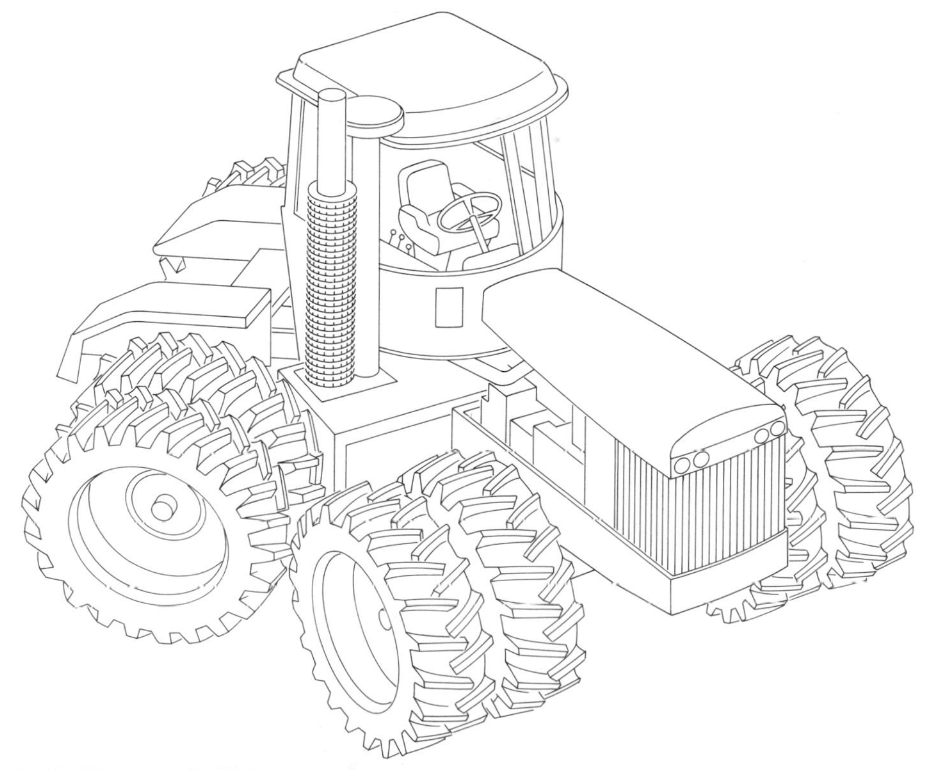

MODERN TRACTORS are exemplified by this four-wheel-drive Deere model, which can develop almost 300 horsepower at an engine speed of 2,100 revolutions per minute. The tractor is articulated: it turns by bending at a center hinge while the wheels remain parallel to the frame. With this arrangement the rear wheels always follow the same path as the front ones.

ground water is limited. Indeed, in some parts of Texas farmers have exhausted the supply and can no longer irrigate their lands. In an attempt to conserve steadily decreasing supplies of available water suitable for irrigation some farmers are turning to "trickle" or "drip" irrigation. Instead of sprinkling or moving the water in open ditches over the surface of the field, pipes deliver small quantities of water directly to the plants. The system is expensive to install, but it reduces evaporation and also cuts sharply the amount of labor needed for irrigation.

The second American agricultural revolution also saw mechanization and related practices extended to animal husbandry, particularly in the production of eggs and broilers, milk, hogs and cattle. Poultry production had led the way earlier, with the development of practical incubators and brooders in the 1870's. The incubators replaced the hen in providing constant warmth to the eggs until they hatched and the brooder provided warmth and shelter for the chicks. In the 1920's it became possible to produce both eggs and broilers in confinement when it was discovered that vitamins added to the diet would help to keep the chickens healthy. After World War II vaccines were developed for several poultry diseases and antibiotics were added to the feed. At about the same time automatic devices for delivering food and water to the confined birds were brought to market. Today virtually all eggs and broilers are produced from poultry confined in highly mechanized quarters.

Dairying keeps the farmer at home because cows must be milked twice a day. Milking machines had been developed by World War I, but they were not widely adopted until after World War II. Since then hand milking has come to an end in commercial dairying. Most dairy barns are also equipped with pipeline systems, eliminating the carrying of the milk. In addition a modern dairy barn is equipped with automatic individual drinking cups, power systems for carrying feed to the cows and automatic barn cleaners, which scrape the gutters behind the cows at the push of a button. Electricity provides the power for the automated dairy barn.

Hogs have been taken out of the traditional mud wallow on many farms. They are raised in pens on easily cleaned concrete floors or wood slats. Water and feed are regularly delivered to each pen mechanically.

The mechanization of cattle feedlots has been one of the most spectacular developments in meat production since World War II. Thousands of cattle to be fattened are brought together in a comparatively small space. Carefully formulated feed is mixed mechanically and delivered to them in mechanized carriers and dispensers, and a constant supply of water is available. Studies have shown that this method of fattening livestock is efficient, but it has certain disadvantages. The runoff from rain falling on the densely populated lots pollutes streams, and the large quantities of manure accumulating in the area can present health hazards. Although efforts are being made to solve these problems, their solution is still in the future.

The process of mechanization is unlikely to be reversed once it gets under way. Mechanization demands better farm management for the efficient utilization of capital-intensive equipment. Because some farms will be managed more efficiently than others there will continue to be widening differences among farms in size, production and in-

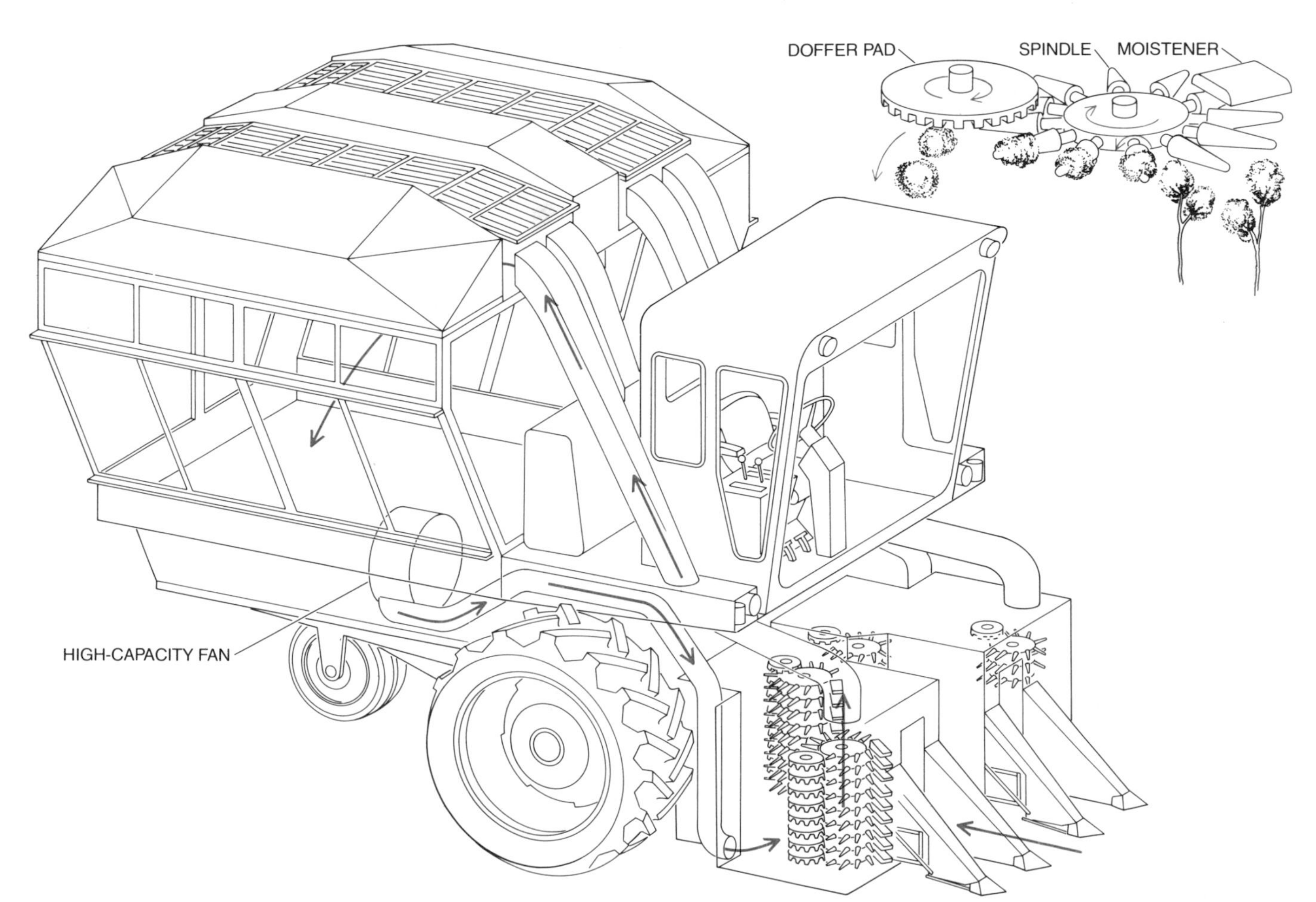

COTTON HARVESTER represents the latest stage in a long process of mechanization in cotton growing. It is a self-propelled machine that pulls cotton fibers from the bolls by wrapping them around moistened spindles, which are then cleaned off by rotating pads called doffers. The enclosed basket behind the cab holds 636 cubic feet of picked cotton. The detail at the upper right shows the mechanism of the spindles and doffers. The machine shown, made by the International Harvester Co., strips two rows of cotton plants at a time.

come. Mechanization has reduced opportunities for employment in farming and may continue to do so.

The range of opinion on the economic and social effects of mechanization in agriculture is reflected in the comments of two writers, Theodore C. Byerly of the Department of Agriculture and Jim Hightower, a critic of the established agricultural-research institutions. A few years ago Byerly wrote: "Continuing development and application of technology in production of food, fiber and forest products can supply the next generation abundantly. It can enable them to take the actions necessary to have clean air, sparkling water and a green and pleasant world in which to live." Some two years later Hightower asserted that "in terms of wasted lives, depleted rural areas, choked cities, poisoned land and maybe poisoned people, mechanization research has been a bad investment."

Perhaps some of the contrast lies merely in the emphasis given, because Byerly also stated: "Our newly applied technology has brought indirect and sometimes unforeseen costs. Pesticides have been dispersed throughout our environment. Crop adjustments [government programs under which land was withdrawn from production in an effort to maintain higher prices] have left people without work or means of self-support. Our abundance has cost the taxpayer in funds for supply management and the producer in depressed prices." And Hightower wrote: "This focus on scientific and business efficiency has led to production (and overproduction) of a bounty of food and fiber products."

The hidden costs of farm mechanization and related changes include the decline in the farm population and in the number of farms, the increase in size and capital needed for a farm to survive, the disappearance of many small towns and the loss of rural social institutions. Otha D. Wearin, an agricultural columnist in Iowa, wrote in 1971: "The productive capacity of power machinery has greatly reduced the farm population. Occupied farming units have become fewer and fewer, and farther and farther apart, as producers with power machinery reach out for more and more land to justify their investment. Country churches, country schools, country society and small country towns have suffered. In fact, many of them have completely disappeared." The loss of some rural institutions has been offset in part by improved transportation, but the full significance of these changes has not been assessed.

Golden Valley County in Montana, where I lived on a ranch until my family's business was wiped out by the Depression, illustrates some of these problems. It is still rural, with no industry and with almost no corporation farming (except for family enterprises that are legally classified as corporations). In 1925, the first year for which census data are available, the county had 492 farms with an average size of 637 acres. In 1978 there were 139 farms averaging 4,693 acres. Thus the area under cultivation had doubled, whereas the number of farms had been reduced by two-thirds. The county had 2,126 inhabitants in 1930 and 931 in 1970. The small increase to 1,026 in 1980 was attributable to the founding of a controversial religious colony. In 1925 the county had two banks, two newspapers, two small hospitals, three practicing physicians, a through railroad line and a creamery. By 1980 all these were gone. One of the three high schools had been closed. There is no extension agent from the state agricultural college. The county

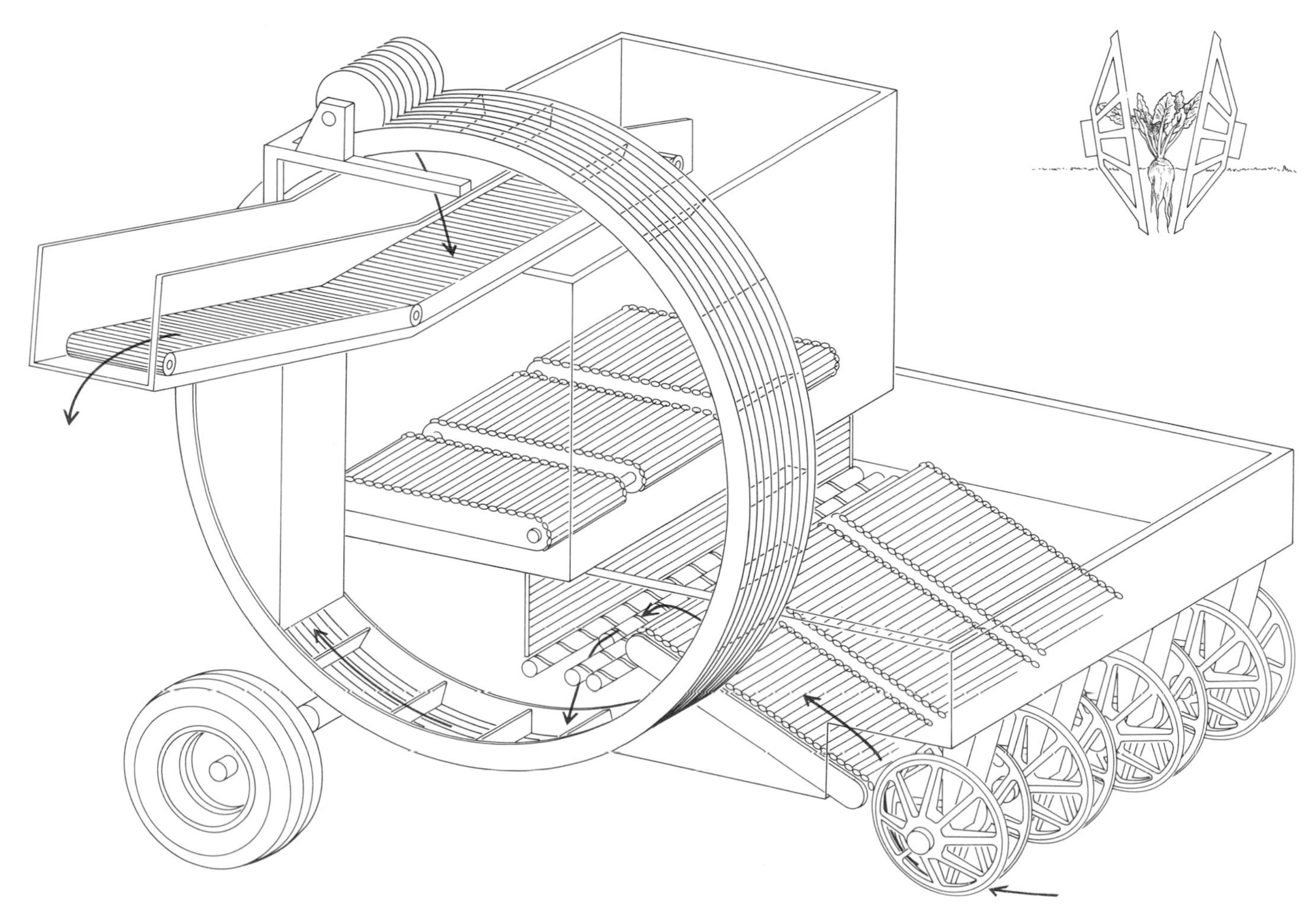

SUGAR-BEET HARVESTER digs beets out of six rows with a set of digging wheels, cleans them on the cleaning bed and lifts them on the large wheel into a holding tank. This is a Deere harvester. Until after World War II the sugar-beet industry relied mostly on hand labor, particularly for thinning beets in the spring and harvesting them in the fall. Now, with machines for planting, thinning and harvesting, sugar-beet production in the U.S. is fully mechanized. Monogerm seeds, each of which produces a single plant, helped the process.

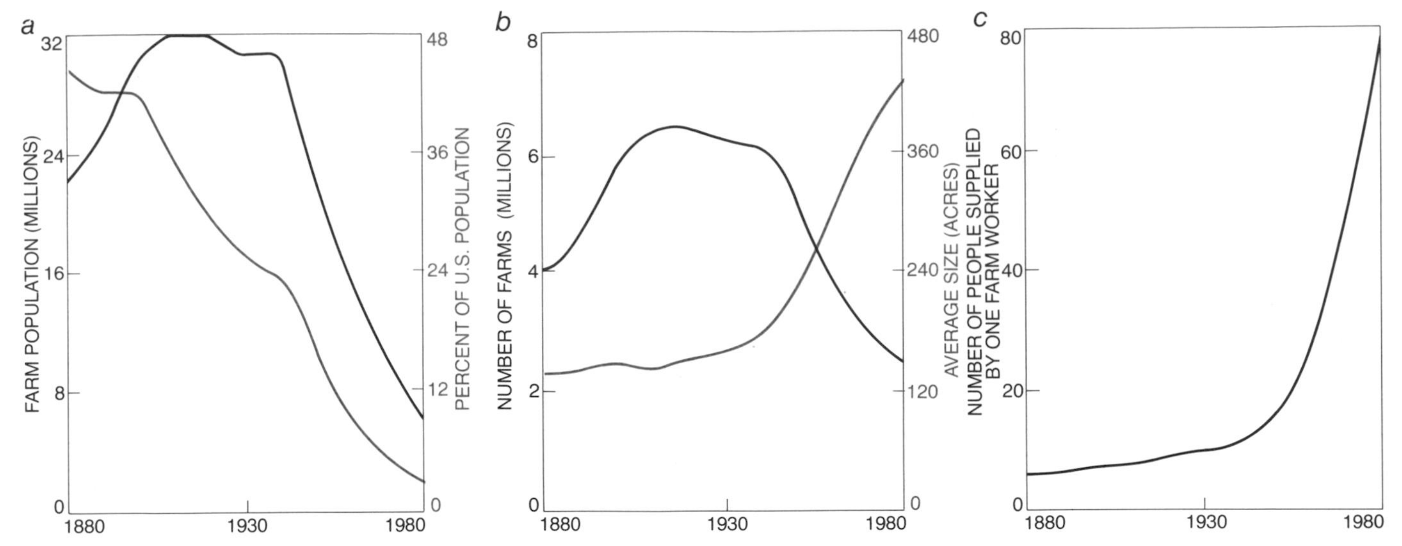

GAINS IN PRODUCTIVITY in agriculture and corresponding declines in farm manpower are charted from 1880 for the U.S. farm population (*a*), the number of farms and their size (*b*) and the number of people supplied with food and fiber by each worker on a farm (*c*).

has yet to acquire a radio or television station. On the other hand, the roads have been much improved since 1930.

Perhaps the major economic beneficiary of farm mechanization is the consumer of agricultural products. Expenditures for food as a percentage of consumer income have been declining since World War II and are substantially lower in the U.S. than they are almost anywhere else in the world. Foreign consumers too have come to rely on the American farmer.

Farmers have seen their disposable income increase from $840 per person in 1950 to $6,553 in 1980. For nonfarm people the corresponding amounts are $1,455 and $8,042. In 1950, however, about 31 percent of farm income came from such nonfarm sources as teaching school, driving a bus, selling insurance and working in a store or a factory. By 1980 the share of farm income from nonfarm activities had risen to 63 percent. Farm income from farming has generally not kept pace with increases in productivity.

Mechanization has been the key reason (but not the only one) for the increase in total production and the higher productivity per man-year that have characterized agriculture in the U.S. The output of wheat has risen from 314 million bushels in 1875 to 669 million in 1925 and to 2.4 billion in 1980. Since 1950 the yield of wheat has increased from 16.5 to 33 bushels per acre, of corn from 38 to 91, of soybeans from 22 to 27. The yield of potatoes has risen from 153 hundredweight per acre to 261. The output per man-year in agriculture has improved at a rate of nearly 6 percent per year, compared with 2.5 percent for all other industries.

The number of farms in the U.S. declined from 6.5 million in 1920 to 5.6 million in 1950 and to 2.4 million in 1980. By now the subsistence farmer has almost disappeared from the countryside except for people who have taken up that way of life not out of necessity but by choice. Today's farmer produces for the market rather than for home consumption.

The farm population and the number of farm workers have declined with the decrease in the number of farms. In 1950 there were 23 million people on farms and 9.9 million farm workers; in 1980 the corresponding numbers were six million and 3.7 million. An important question is what has happened to the displaced farm population. In earlier times surplus farm workers supplied much of the manpower needed for the industrialization of the economy. Since the second American agricultural revolution many of the displaced farm people have gone to the cities, often to an impoverished and bleak situation. Others are now among the rural poor.

Another important question is whether or not American society should tolerate or even encourage the persistent erosion of the farm population. Mechanization has helped to keep the family farm viable by enabling the family to work the acreage needed for an economically sound unit. On the other hand, the investment now required to establish a new farm makes it almost impossible for a young family to enter farming unless they can continue on a farm that is already in the family. Even the established farmer finds it difficult to finance the purchase of additional land and new equipment.

Who should be responsible for planning for the workers who might be displaced by new technologies? Certainly not the inventors and technologists. The country cannot continue to maintain agricultural production at a satisfactory level if research is halted. Henry A. Wallace addressed this question in 1940, when he was Secretary of Agriculture and when research was under attack as adding to farm surpluses. "Science, of course, is not like wheat or automobiles," he said. "It cannot be overproduced.... In fact, the latest knowledge is usually the best. Moreover, knowledge grows or dies. It cannot live in cold storage. It is perishable and must be constantly renewed."

Nor should the manufacturers and suppliers of agricultural equipment be expected to decide whether or not a particular technology should be made available if it might force some people out of work. The Rust brothers found that such restraints did not work with their cotton picker. The answer is not to limit research or stop the production and use of new machines, even though some changes in emphasis may be appropriate for research done with public funds. Farmers and their families today live longer and healthier lives at least in part because of the mechanization of American agriculture. Whatever solution may be found to the problem of disemployed farm workers, it would hardly be humane to return them to dawn-to-dusk labor chopping cotton, thinning beets or flailing grain.

In the early 1980's there are some indications that agricultural productivity may have reached a plateau. At least it is not increasing at the rate that prevailed from 1950 to 1980. The number of commercial farms is still declining, but the rate of decline is lower than it was over the past 30 years. Approximately the same total acreage is being farmed. What is most surprising, the rural population has increased faster than the urban population in the past 10 years. Many of the people who have left the cities are actually living a suburban life in areas formally classified as rural, but there may have been a slight increase in the number of people living on farms.

With one exception, no major further advances in the mechanization of agriculture are now in sight. Indeed, if the future can be judged by the past, no major change will take place until a new source of power comparable to the horse and the internal-combustion engine is adapted to agriculture and adopted by farmers. One can expect workers in agricultural research and manufacturers to continue advancing the technology by improving old machines and designing new ones, particularly machines that reduce the demand for human labor and those that substitute technology for land. These achievements will depend in part on a revival of agricultural research, which for several years has felt the effects of declining expenditures in both the public and the private sectors.

The one exception is the application of computers to farm management, which is already under way and seems likely to have a powerful influence. The computer should lead to more efficient management of machines and energy and should help in other farm operations such as cost accounting, mixing feed and deploying fertilizers and other resources efficiently. Some farmers now have computers of their own, and many others have access to computer systems through their county agricultural agent. Packages of computer programs have been developed to meet the needs of farmers in particular states and communities, and one major computer manufacturer is introducing hardware and software designed to serve small farms.

What might a mechanized American farm be like in, say, 2020? First, it will be a large, commercial enterprise operated by a family. It is becoming increasingly evident that the family farm has several advantages over the corporate farm, particularly because the family provides all or most of the labor, directly supervises the hired workers and has the possibility of deferring profits. Second, the internal-combustion engine will still be the primary source of power unless solar energy has been effectively harnessed. Third, additional mechanization will have further reduced farm drudgery. Machines and automated systems may be available for harvesting fruits and vegetables that in 1982 are still picked by hand, for transplanting plants, for adjusting the flow of irrigation water to individual fields and perhaps even for mending fences. Fourth, the extensive application of computers will have made the farm considerably more efficient. Finally, the farmer and his family will enjoy virtually all the amenities of the city-dweller while still having the feeling of being a group apart that works with soil, water and seed to provide large numbers of other people with a plentiful supply of food.

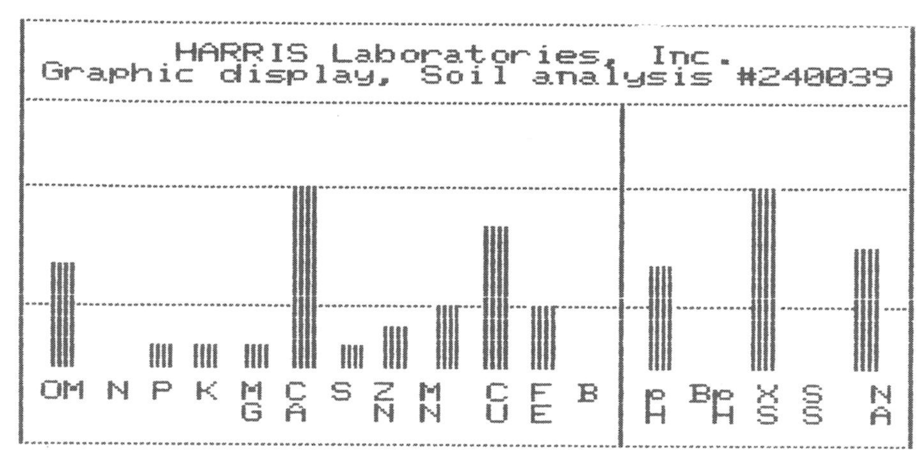

GROWER = MENDONCA DAIRY
SAMPLE ID = F1-S1 12-23-81
2.5 BALES COTTON

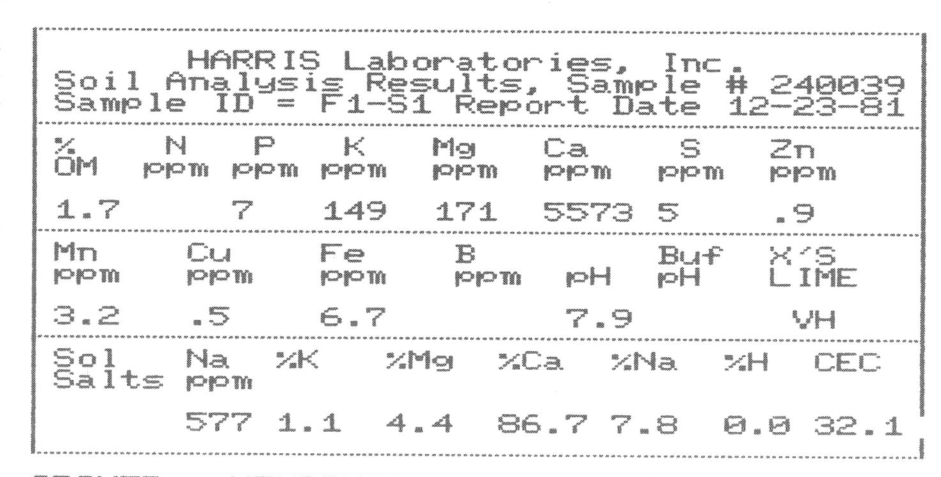

HARRIS Laboratories, Inc.
Soil Analysis Results, Sample # 240039
Sample ID = F1-S1 Report Date 12-23-81

% OM	N ppm	P ppm	K ppm	Mg ppm	Ca ppm	S ppm	Zn ppm
1.7		7	149	171	5573	5	.9

Mn ppm	Cu ppm	Fe ppm	B ppm	pH	Buf pH	X'S LIME
3.2	.5	6.7		7.9		VH

Sol Salts	Na ppm	%K	%Mg	%Ca	%Na	%H	CEC
	577	1.1	4.4	86.7	7.8	0.0	32.1

GROWER = MENDONCA DAIRY
2.5 BALES COTTON
SEE SPECIAL TEST RESULTS

HARRIS Laboratories, Inc.
Fertilizer Recommendations for #240039

Your ID#	Plant Food Needs (in lbs./acre)
F1-S1	For 2.5 BALES COTTON

SULF LBS	N	P205	P205 Build	K20	K20 Build	MgO
360	125	95	65	80		30

Acres	S	Zn	Mn	Cu	Fe	B
30	38	9	6	----	7	

GROWER = MENDONCA DAIRY
REPORT DATE = 12-23-81
SOILDATA RECORD NO. = 9

ADVENT OF COMPUTERS in agriculture, which is just beginning, holds the promise of greater efficiency in managing machines and energy and in operations such as cost accounting, deploying fertilizer and mixing feed. A printout like the one shown can be transmitted to a farmer's computer terminal after he has sent a sample of soil to a laboratory to be analyzed. *OM* in the analysis means organic matter; most of the other amounts are in parts per million.

3

THE MECHANIZATION OF MINING

The Mechanization of Mining

by ROBERT L. MAROVELLI and JOHN M. KARHNAK

Today more than 80 percent of the mineral needs of the U.S. economy are met by only 1 percent of the labor force. Here the mechanization of mining is examined in terms of its effect on the mining of coal

Mining has long been perceived as hard, dirty and dangerous work. It is also the preeminent materials-handling industry, and its products are extracted from the earth in enormous quantities. The work was once done with pick and shovel, and gains in productivity were achieved by increasing the tonnage of material one man could dig in a working day. Now the capabilities of the individual miner are greatly amplified by machines and systems of machines. The changes in technology not only have raised productivity but also have had a pronounced effect on the health and safety of miners.

Various energy sources have long been exploited to augment human labor in mining. Animals provided power to operate pumps for mine drainage and bellows for ventilation. They also transported ore horizontally through the underground workings and hoisted it to the surface. Water power, applied through ingenious systems of gears, spindles and shafts, often took over some of the same tasks. Thermal energy from wood fires was employed to heat rock surfaces, which subsequently shattered when they were chilled with water. This technology was described in detail in 1556 by Georgius Agricola in his treatise *De Re Metallica*.

The development of the steam engine early in the 18th century set the stage for the true mechanization of mining. For the first time the mining of coal and ores could be carried to great depth and horizontal extent with enough power for drilling, cutting, loading, hauling and hoisting as well as for pumping water from the mine and providing adequate ventilation. Mines are not factories, however, and advances in productivity tend to come slowly and at high cost. Many promising machines have proved inadequate and unreliable under actual mining conditions.

The greatest gains in mine mechanization came in a burst of innovation following World War II, when robust new machines were developed, along with new methods for deploying them. It is a tribute to such mechanization that today more than 80 percent of the mineral needs of the U.S. economy can be met from domestic sources with the work of less than 1 percent of the labor force.

A true measurement of productivity calls for a complex analysis of capital and labor costs; in mining geological factors such as mineral quality and accessibility must also be considered. Rapid and often unpredictable changes in these factors further complicate the analysis. Nevertheless, productivity expressed in terms of yield per unit of labor input is an important measure of overall mining efficiency and is the one we shall use in this article.

The effects of mechanization on productivity and on working conditions are deeply intertwined. In coal mining, which is by far the largest segment of the mining industry in the U.S., productivity showed only small improvement until 1950. It then climbed sharply, reaching a peak in 1970 from which it has since fallen back (for reasons we shall describe). Reliance on human muscular exertion has been greatly reduced. Fatalities in mine accidents, which routinely exceeded 2,000 per year until 1930, have been cut by a factor of more than 10. Over this period, however, some hazards to the underground miner's life and health were actually increased by new machinery. A shift to surface mining, or strip mining, since World War II has helped greatly to reduce the number of mine fatalities both because surface mining is inherently safer than underground mining and because the higher productivity of surface mining has lowered the number of workers at risk.

In 1925 some 588,000 men, about 1.3 percent of the nation's labor force, were needed to mine 520 million tons of bituminous coal and lignite, almost all of it from underground. Last year production was up to 774 million tons, but the work force had been reduced to 208,000. Furthermore, only 136,000 of that number were employed in underground mining operations. The highly mechanized and highly productive surface mines, with just 72,000 workers, produced 482 million tons, or 62 percent of the total.

Last year approximately 765,000 men and women were broadly classified as miners, but 300,000 of them, or roughly 40 percent, worked in oil and gas fields and so were miners by definition only. The 208,000 coal miners constituted roughly half of the remainder; another 93,000 were engaged in nonmetal mining and quarrying, and some 57,000 mined a variety of metallic ores (chiefly iron and copper). In addition 107,000 were employed in mills and shops associated with mines; these employees too are classified as miners, although office

BUCKET-WHEEL EXCAVATOR exemplifies the large and costly machines that have been introduced into coal mining in recent decades. The excavator is one of two such machines at the Captain Mine in southwestern Illinois. The mine is owned by the Arch Mineral Corporation and has an annual output of 4.5 million tons of coal. One of the excavators removes the first one or two feet of topsoil; the machine shown strips off the next 15 to 20 feet of earth. The material removed by the two machines is carried more than 8,000 feet by belt conveyors and held in separate piles for eventual restoration of the mined-out land. (State and Federal regulations require that the land be restored to its approximate original contours and, if it was farmland, that there be no loss of agricultural value.) Still more overburden is excavated by shovels and draglines before the first coal seam is reached some 85 feet below the surface. Another 24 feet of earth is then stripped off to gain access to a second seam. The two seams have a total thickness of about 10 feet. Almost 60 percent of the man-hours at the mine are spent in removing the overburden and reclaiming the land and only about 40 percent in mining the coal itself. The excavator is built in West Germany by O & K Orenstein & Koppel Aktiengesellschaft.

workers are not. This work force satisfied 83 percent of the nation's demand for fossil fuels on an energy basis and nearly 90 percent of the demand for all other minerals on a dollar basis, the only common denominator for that diverse category. (Of the 56.4×10^{15} British thermal units of energy derived from domestic fossil fuels in 1981, coal supplied 28.4 percent, petroleum 36.6 percent and natural gas 35.0 percent. The nation's expenditure for all fossil fuels last year was $260 billion, including $77 billion for imported oil; the value of coal at the mine was about $20 billion.)

Last year's production of 774 million tons of bituminous coal and lignite exceeded preliminary estimates of the combined tonnages of the next three minerals in greatest demand: iron ore (246 million tons), copper ore (306 million tons) and phosphate rock (203 million tons). Of materials removed from the earth coal was exceeded in tonnage only by crushed and broken stone (873 million tons). If account were taken of the huge mass of earth and rock that had to be removed to gain access to the 482 million tons of coal produced by surface mines last year (estimated at 10 tons of overburden for each ton of coal), the coal industry would far surpass all other industries in mass of material handled.

Today few minerals are mined primarily underground and most of those are of low tonnage. The principal metal-

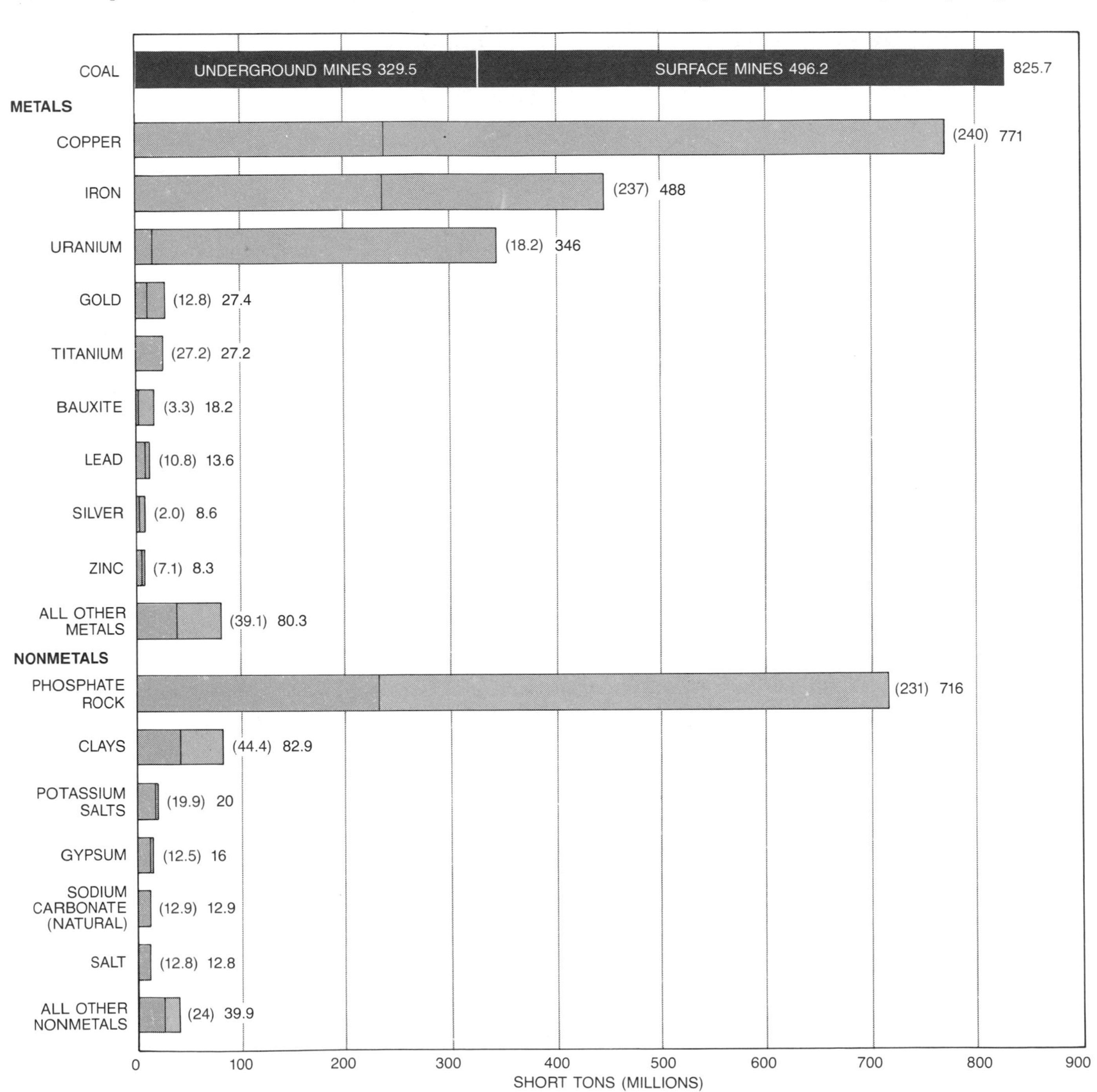

OUTPUT OF U.S. MINING INDUSTRY is dominated by coal: 825,700,000 tons of bituminous coal and lignite in 1980, the most recent year for which a comparison with other mineral products can be made. The bars for the other minerals show the total tonnage of material handled, including waste (*gray*), and the net amount of usable crude ore recovered (*color*). Although comparable data are not collected for coal, the tonnage of waste material handled in surface mining, which accounted for 60.1 percent of the total production, probably exceeded 4,000 million tons. This unrecorded tonnage represents the great overburden of earth and rock that had to be removed to get at coal seams lying as much as 200 feet below the surface. Not shown are two high-volume but low-value products of the mining industry: sand and gravel (794 million tons) and crushed and broken stone (1,060 million tons). Compared with coal, the 1980 domestic production of petroleum was about 570 million tons (3.7 billion barrels) and of natural gas 450 million tons (20.1 trillion cubic feet).

lic ores obtained entirely or in large part from underground sources are antimony, lead and tungsten (from 98 to 100 percent underground), molybdenum (62 percent) and silver (60 percent). Among nonmetals only three major substances are extracted primarily by underground mining: potassium salts, sodium chloride and natural sodium carbonate. The number of American miners who spend their working days belowground does not exceed 160,000, and about three-fourths of these are coal miners. The variety and the quantity of materials obtained by surface mining are much greater. About 95 percent of all metallic ores and 75 percent of all crude nonmetallic ores come from surface mines.

In the surface mining of iron and copper ore improvements can be attributed chiefly to economies of scale: larger earth-moving machines, larger drill holes for blasting and larger shovels for loading larger trucks. In describing productivity in the mining of iron ore, however, one must distinguish between the output of crude ore and the output of usable ore. In copper mining the distinction is made between tons of crude ore and tons of recoverable metal. In both cases greater productivity in the mining of crude ore has been partially offset by a decline in the metal content of the crude material. Since 1952 the output of crude iron ore in long tons per man-year (the traditional measure in that industry) nearly quadrupled from about 3,600 to 12,700. Over the same period the output of usable ore increased about 50 percent, from 2,750 to 4,200 long tons per man-year. The peak of 5,200 long tons came in 1972.

Productivity gains in copper mining over the same period were substantially lower. Between 1950 and 1981 the output of crude ore per man-year increased from about 3,700 short tons to a little more than 9,000, but the output of recoverable metal rose at a much lower rate, from 35 tons to 50 tons per man-year. In 1950 each ton of copper ore mined yielded an average of 19 pounds of copper; by 1981 the average yield had dropped to about 11 pounds per ton.

In open-pit copper mining there are three basic operations: drilling blast holes, loading trucks and hauling the ore. An industry analysis has shown that the smallest productivity increase has been achieved in hauling, even though trucks of much greater capacity are now in operation. The reason is that as the pit is made larger and deeper the ore must be hauled and lifted farther. It appears that economies of scale have reached a temporary plateau in open-pit copper mining. The industry expects that future gains can be made with the aid of computers to improve the coordination of the mining system, by the use of self-propelled scrapers and crushers and by

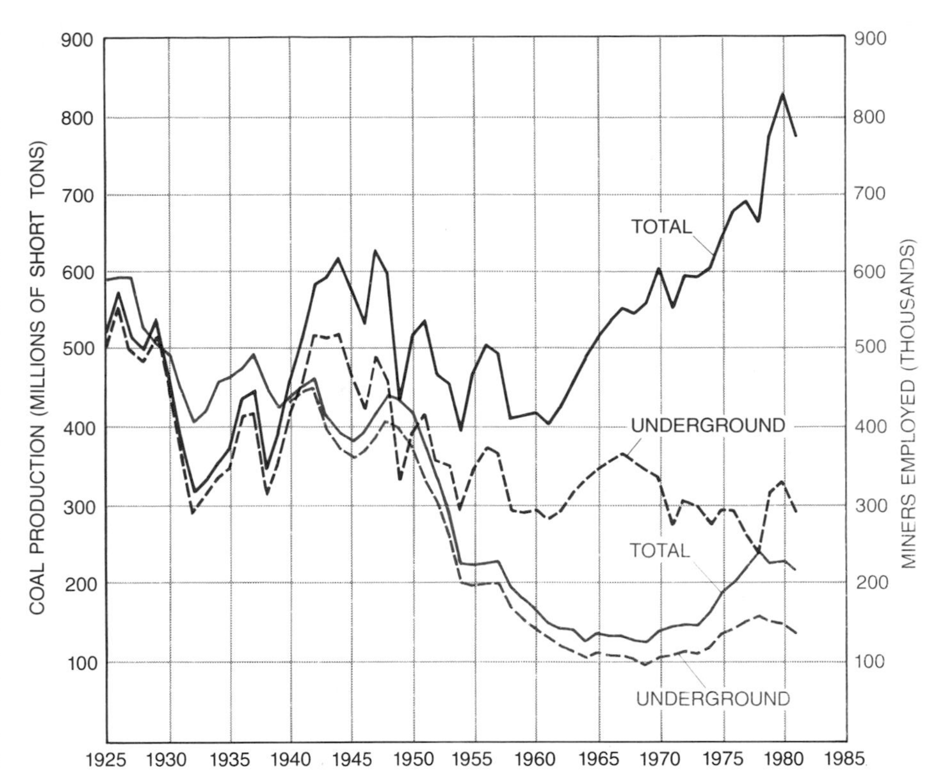

HISTORY OF COAL MINING in the U.S. since 1925 is marked by sharp swings in demand and in the size of the work force (*color*). The biggest change, however, has been the shift away from underground mining since World War II. As late as 1945 some 94 percent of the nation's 383,000 miners worked in underground mines. Today there are only about 208,000 miners, and fewer than two-thirds of them labor underground. In the late 1920's the average miner produced about 950 tons of coal per year. The present output is about 3,700 tons per man-year.

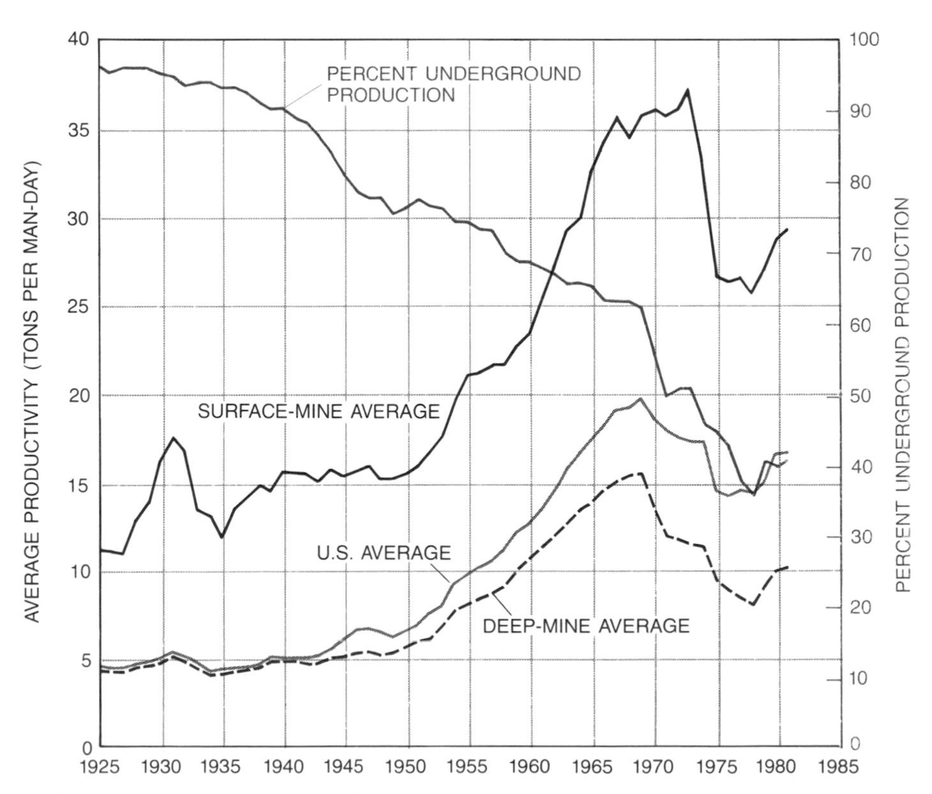

PRODUCTIVITY PER MAN-DAY in coal mines crept upward at a rate of only 1.6 percent per year between 1925 and 1950. In the next 20 years, however, with the introduction of continuous mining machines belowground and of larger earth-moving machinery in surface mines, output per man-day climbed at an annual rate of 5.3 percent. Some of the productivity gains have since been lost, partly because of Federal legislation designed to improve miners' health and safety. After 1974 surface-mine productivity fell sharply for a number of reasons, including land-reclamation laws. In the past three years productivity has begun to rise again.

adopting belt conveyors for difficult long hauls.

Mechanization in coal mining can be traced almost to the beginning of the industry. Some of the first steam engines were put to work pumping water from mines, and they were later applied to mine ventilation and the transport of men and materials from the surface to the mine-working level. A few steam-powered drills and haulage locomotives were used underground, but they were soon replaced by devices driven by compressed air. Beginning in 1888 electricity became available in the mines, supplying light and power for machinery. By the early 1900's power equipment was commonplace in American coal mines for a number of operations, including the drilling of blast holes and the undercutting of the coal face in preparation for blasting. By the end of World War II 90 percent of the coal mined in the U.S. was undercut by machine.

In spite of early progress in the mechanization of underground coal mining, the improvement in productivity in the first half of the 20th century was far from dramatic. In 1897 a man with a pick and shovel could mine three tons of coal per day. In 1925 the average miner's output in a somewhat shorter working day was 4.5 tons. By 1945 productivity had increased by only a little more than 10 percent, to five tons per man-day. In comparison the gains since then have been remarkable. In 1969 productivity briefly reached 15.6 tons of coal per man-day, and the work force required by underground mining operations had shrunk to 99,000.

The decline in productivity in the 1970's can be attributed to several factors. The low point of 8.4 tons per man-day came in 1978, and employment in underground mining rose in proportion (to 160,000). One important cause was unquestionably the institution of more stringent safety regulations; another was labor strife. The diminishing quality of the accessible resources also contributed: in all forms of mining the most accessible deposits are worked first, and so mining tends inevitably to become more expensive as it is extended deeper into the earth or to more distant sites. Some mine operators contend that mechanization itself is a third cause of lowered efficiency; reliance on larger but fewer machines, they point out, makes the mining operation more susceptible to interruption as a result of breakdowns. Whatever caused the loss of coal productivity, the recent recovery has brought it up to about 10 tons per man-day.

Strip mining, with its potential for mechanization on a vast scale, grew little until World War II. Between 1925 and 1941 the output of surface mines increased only slowly, from 3.2 percent

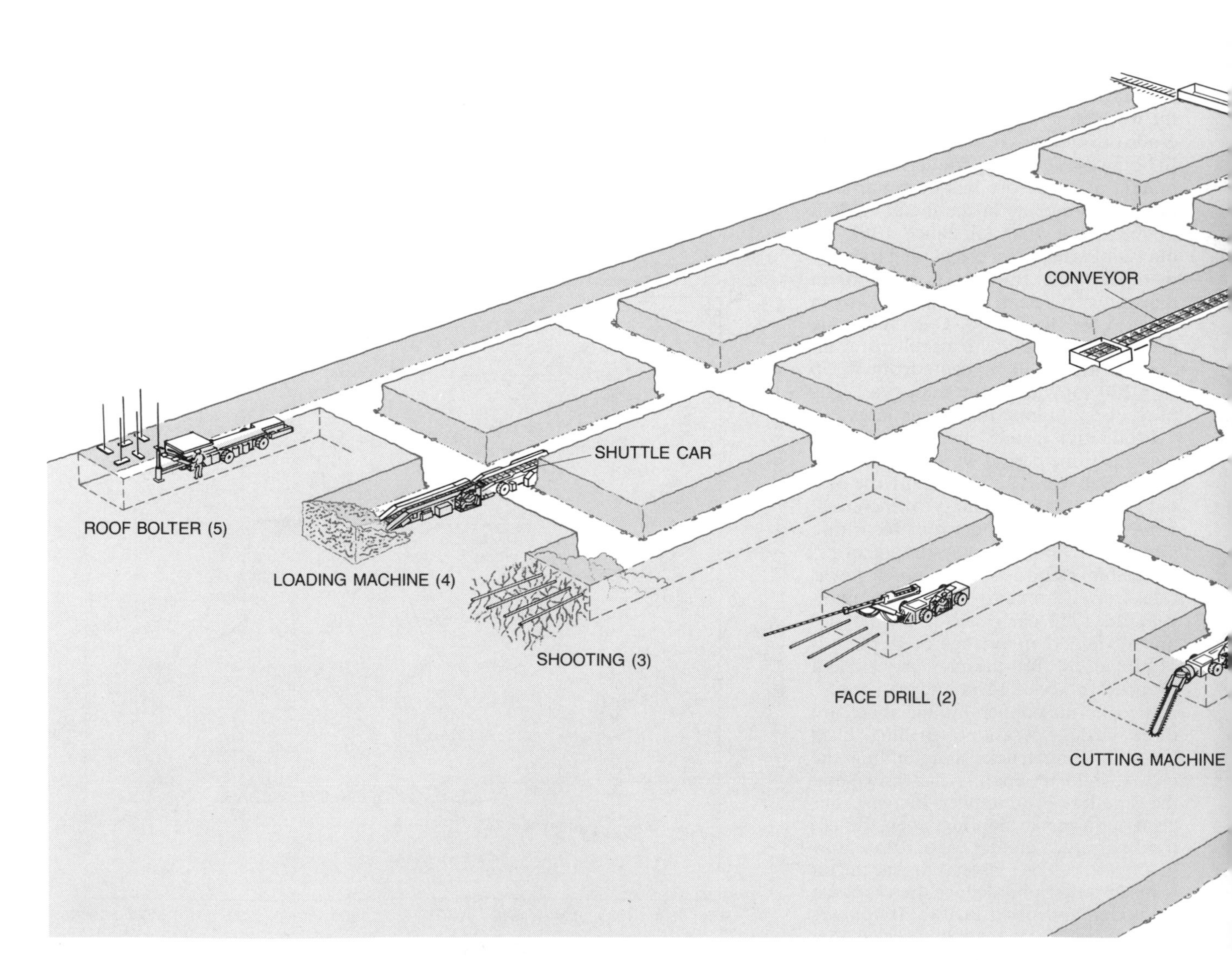

ROOM-AND-PILLAR MINING has been the standard method of underground coal mining in the U.S. since the 19th century. The rooms are empty areas from which coal has been removed; the pillars are blocks of coal from 40 to 80 feet on a side, left standing to support the roof of the mine. In the final stage of the mining of a seam the coal in the pillars can be extracted, allowing the roof to fall. In conventional mining, diagrammed at the left, five operations are carried out in sequence. Typical machines for the operations are shown at the right. In the first step a slot roughly six inches high and 10 feet deep is cut across the base of the seam by a machine with a long cutter

of total coal production to 9.2 percent. By 1945, however, the share of coal coming from surface mines had more than doubled to 19 percent. As noted above, strip mines now account for 62 percent of American coal.

The productivity of surface mining has consistently been higher than that of underground mining. The productivity was about 11 tons per man-day in 1925 and 15.5 tons in 1945. A peak of 36.7 tons per man-day was reached in 1973. As in underground mining, a substantial decline followed. New requirements imposed on mine operators were among the reasons for the decline, although in surface mining the requirements had to do not so much with safety as with the restoration of the mined lands to their approximate original contours and agricultural quality. Productivity fell to less than 26 tons per man-day in 1978, but it has since recovered and is now approaching 30 tons.

Whereas the productivity of surface mining has been paced in large part by increases in the size and efficiency of earth-moving and ore-moving equipment, productivity gains in underground mining have required the introduction of new technology. This has been particularly true of coal mining. One of the principal deterrents to technological innovation has been the geological diversity of coal seams. In West Virginia coal is mined from seams as thin as two feet and as thick as 18 feet. Most of the seams are quite flat and many can be entered directly from the side of a hill. Few West Virginia mines are more than a few hundred feet below the level of surface transportation. In Colorado, in contrast, some mines are 3,000 feet deep and have seams dipping 30 degrees from the horizontal. There is also diversity in the size of mines, with output ranging from a few thousand tons per year to more than 15 million.

Because coal deposits are widespread in the U.S. coal mining has attracted thousands of small operators. Between World War I and World War II the number of bituminous mines ranged from about 5,400 to more than 9,300, with virtually as many mining companies. Today some 3,500 companies operate about 6,000 mines. The largest company, the Peabody Coal Company, mines no more than about 8 percent of the total; it takes 50 of the largest companies to mine 65 percent.

The typical underground coal mine in the U.S. is laid out in a checkerboard of rooms and pillars, a mining method dating back to the 19th century. The rooms are empty spaces from which coal has been removed. The pillars are blocks of coal from 40 to 80 feet on a side left standing to support the roof of the mine. The deeper the mine is, the larger the pillars must be. Eventually

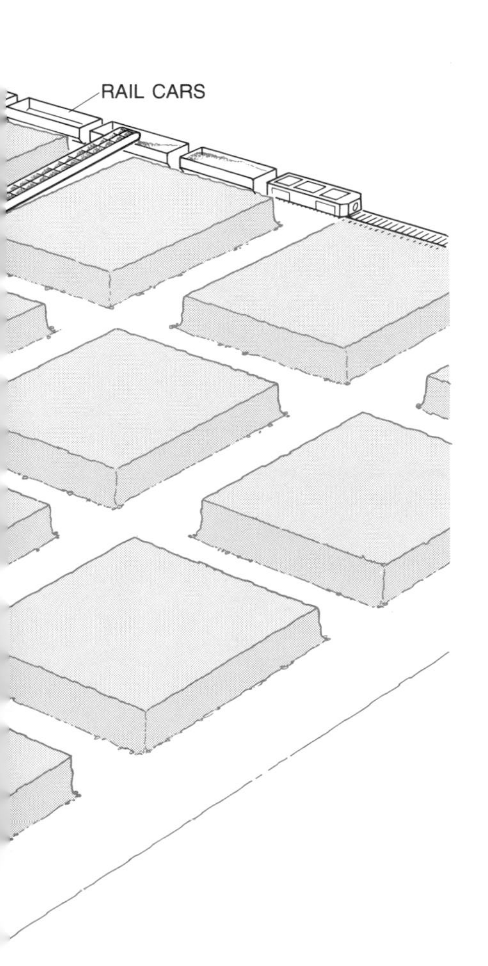

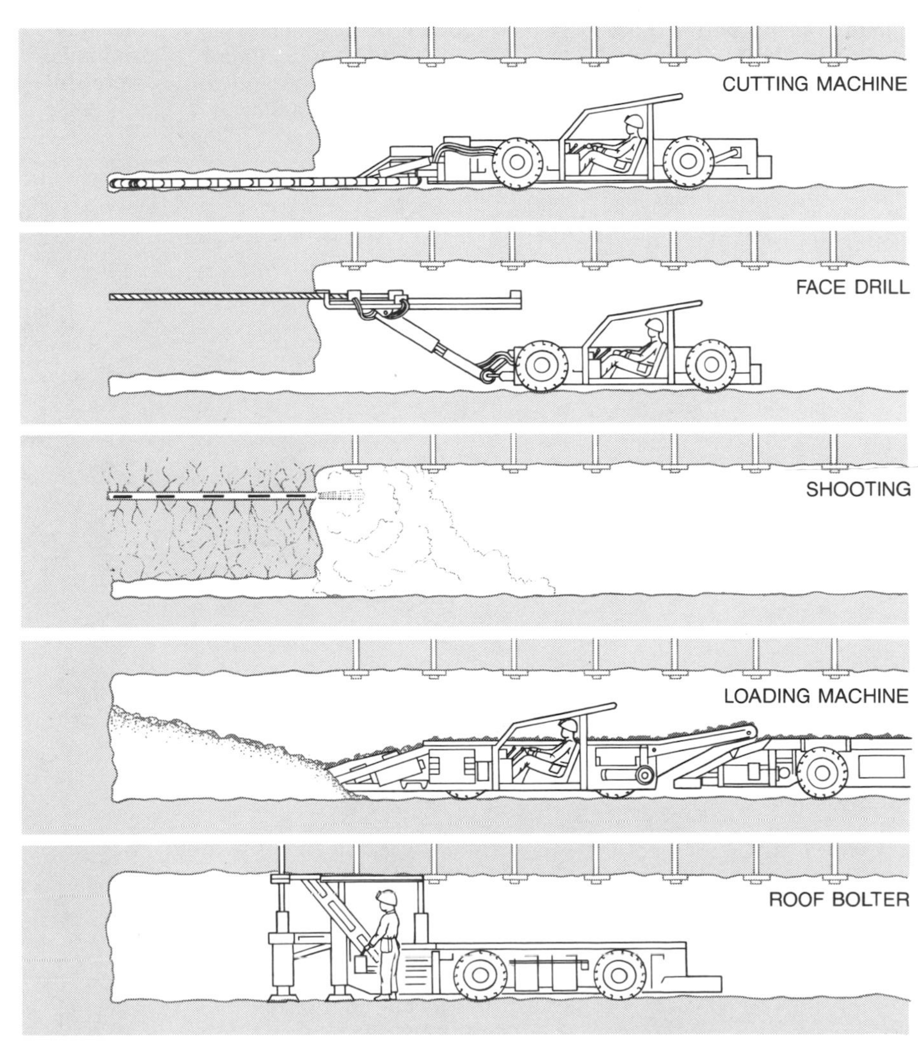

bar. A series of holes, also about 10 feet deep, are then drilled in the face of the coal above the slot and are loaded with explosives. The detonation of from 10 to 15 pounds of explosives fractures up to 50 tons of coal, which spills out onto the floor of the mine. A loading machine conveys the shattered coal into a waiting shuttle car, which hauls the coal to a belt conveyor. The belt in turn carries the coal to the main haulage line (which can be another belt or a rail line) for transport from the mine. The final step is to insert a series of long steel bolts into the roof in order to bind the overlying layers of shale into a strong laminate. The sequence of operations is then repeated.

the coal in the pillars can be recovered by mining the pillars at the farthest point of advance and retreating to the mine entrance, letting the roof collapse in a controlled way as pillar after pillar is salvaged.

A room-and-pillar mine is opened by excavating entry and exit tunnels in the virgin coal. Miners and machines are transported to the working faces of the mine by an underground railroad, which may also haul away the coal. The mining proceeds from multiple "entries" cut parallel to the main haulage lane and reached by cross tunnels. The sequence of mining operations at the coal face in 1947 would have been familiar to a miner of 1897, except for the replacement of muscle power by machines.

In conventional mining the energy needed to dislodge coal from the seam is supplied by chemical explosives. In the 1940's American coal mines annually consumed almost 500 million pounds of explosives, about half of the amount used in all mining and quarrying and some 40 percent of all explosives sold for industrial purposes. (Last year the coal industry used an estimated 2.25 billion pounds of explosives, which was half of the industrial total; 98 percent of this amount, however, was consumed in surface mines.) To make the blast effective a slot six to eight inches high and about 10 feet deep is first cut in the base of the seam. The undercutting is now done by a machine equipped with a movable cutter bar that resembles a chain saw; in 1897 a miner used a pick to hack out a smaller slot. The slot provides an additional free face into which the coal can expand when it is blasted.

The next step is to drill a series of holes into the face of the coal parallel to and above the slot. The holes are two inches in diameter and 10 feet deep, and they are spaced about two feet apart. Each hole is loaded with between seven and nine sticks of explosive. Ten to 15 pounds of explosive will shatter up to 50 tons of coal. The explosive charge is wired with electric blasting caps and detonated from a safe distance. The resulting pile of shattered coal is scooped up from the mine floor by a loading machine with two crablike arms and conveyed to a waiting shuttle car or other haulage device. By 1947 some 60 percent of underground coal was mechanically loaded. More than 10,000 animals, however, most of them mules, were still employed for haulage.

The final step in room-and-pillar mining is to support the newly exposed roof. In 1947 the standard roof support was a framework of heavy timbers. A year earlier, however, a few coal mines had begun experimenting with roof bolts, which had been introduced as early as 1927 in the mining of minerals other than coal. The steel bolts are commonly four to six feet long, and they are generally inserted at intervals of every four feet in a gridlike pattern. The bolts are held in place either by a mechanical expansion shell or, more recently, by polyester resin. The bolts bond the various rock strata overlying the coal seam to support the mine roof. It was by these methods that the average underground mine of 1947 achieved an output of five tons of coal per man-day. The most fully mechanized mines, with seams up to 11 feet thick, attained a productivity twice as great.

By late 1948 the first models of a continuous mining machine were being secretly tested in several mines. The machine, which was built by the Joy Manufacturing Company of Pittsburgh, removed and loaded coal from the solid face in one step. The original machine

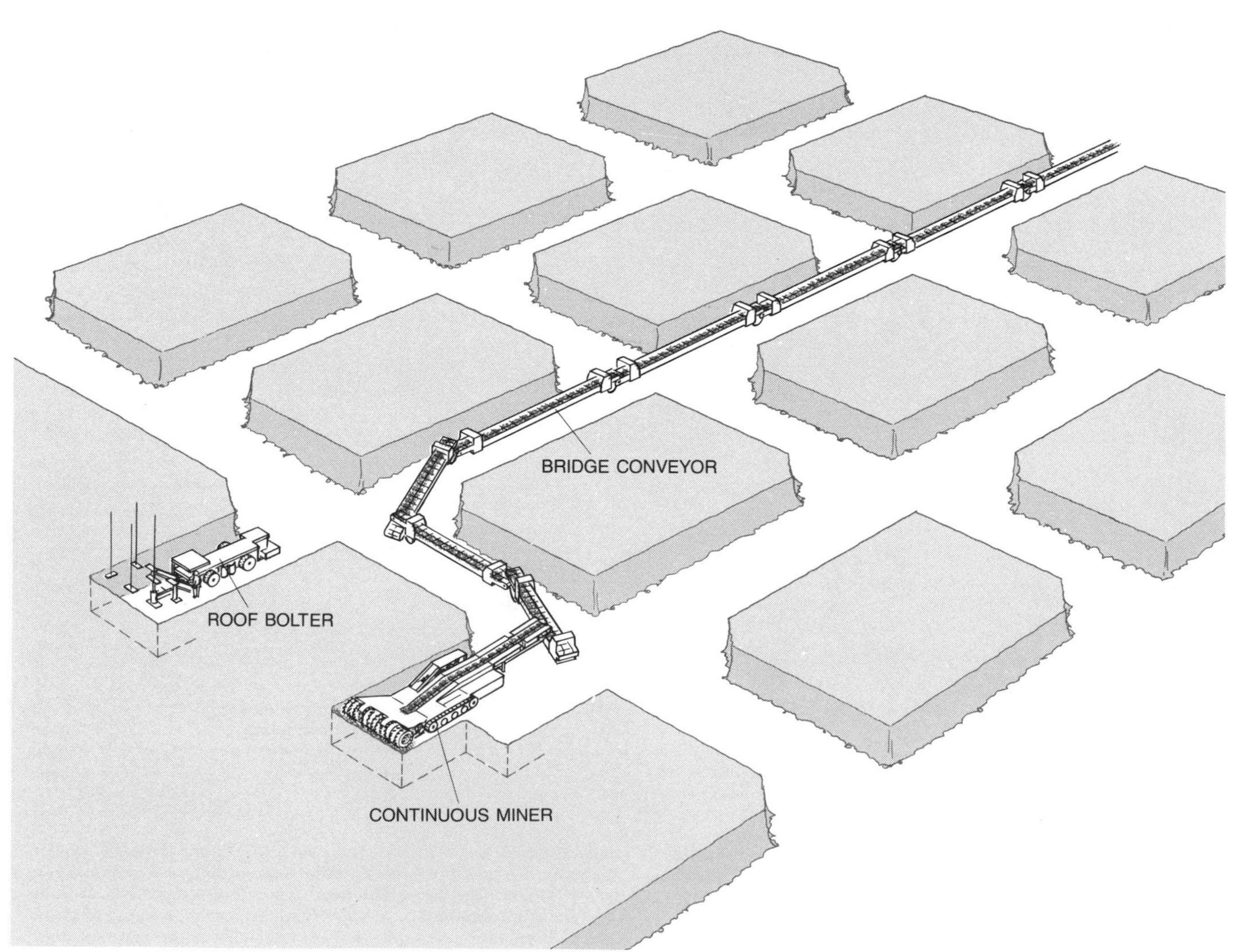

had a gang of cutter chains mounted vertically on a swiveling head; the chains were driven into the base of the coal face and then ripped the seam upward from the floor to the roof. The coal was carried to the rear of the machine by a conveyor and dropped into a waiting shuttle car. The Joy company stated that its machine could raise productivity in a typical mine to 15 tons per man-day.

The industry was skeptical. Many of its customers paid a premium of several dollars per ton for coal delivered in sizable chunks. If the new mining machine increased the fraction of small cuttings and "slack" (pieces less than three-eighths of an inch in diameter), most of the incentive for higher productivity would be wiped out. Some mining engineers also questioned the economics of fracturing coal by electrically driven machinery when that step could be done cheaply with explosives. At the time the question seemed so important that a continuous miner expressly designed to minimize the cost of fracturing coal was financed by an industry-supported organization, Bituminous Coal Research, Inc. The machine was unsuccessful. As improved models of the Joy miner and competitive machines were developed, the industry's skepticism slowly abated.

Continuous mining machines being made now are larger and more powerful than the original Joy machine but work on the same principle. The main difference is that the cutter chains have been replaced by a rotating drum about two feet in diameter and from eight to 16 feet wide; it is fitted with steel bits tipped with silicon carbide cutting edges. The drum turns at about 60 revolutions per minute. It is driven into the top of the coal seam a distance nearly equal to its diameter and is then moved downward, shattering the coal. The fractured coal is pulled to the center of the face by gathering arms, then transferred by a conveyor to a waiting shuttle car or to another conveyor. When the machine has advanced 20 feet or so, it is moved to another place and the exposed roof is bolted. Continuous mining machines were widely adopted in the late 1950's and the 1960's, and now almost two-thirds of all underground tonnage is extracted in this way.

As coal mining is extended to greater depths, larger pillars must be left to support the overlying strata. Although overall coal extraction can be raised from about 50 percent to 70 percent by removing the pillars in reverse sequence and allowing the roof to collapse, such retreat mining requires a great deal of skill and experience, and it can be more dangerous than other forms of mining. As a result an alternative to room-and-pillar mining has received increasing attention. It is known as longwall mining, and it has been practiced for many years in deep European mines.

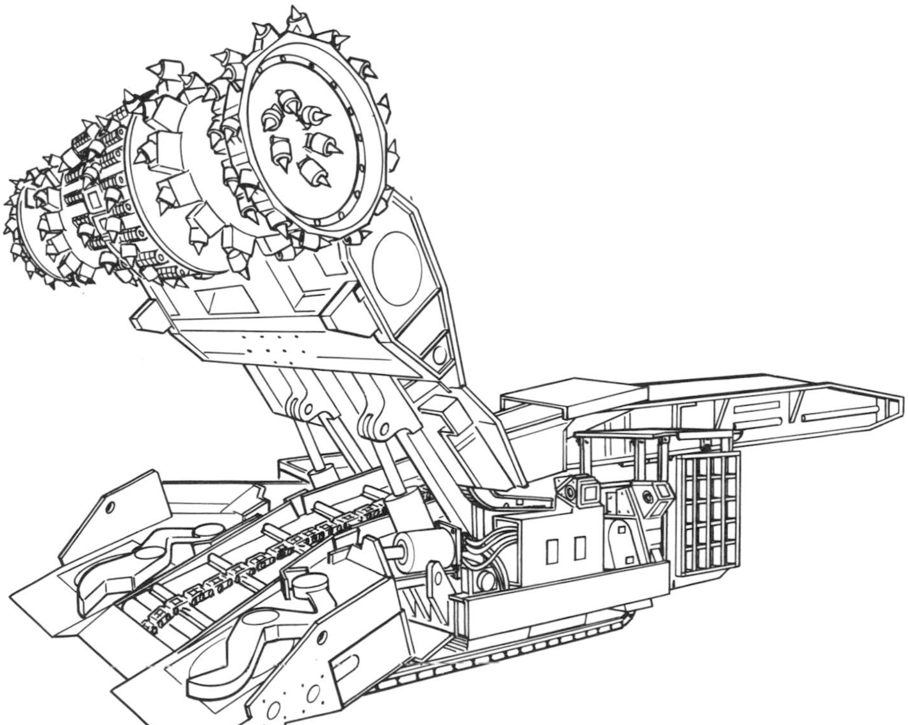

Longwall mining is a technology for extracting a continuous block of coal. The block is usually from 400 to 600 feet across the face, and it can be as much as a mile long. A specialized machine, either a plow or a shearer, travels along the face on a guide or track, cutting the coal and depositing it on a conveyor that carries it out to a main haulage junction at the end of the face. An essential element of the longwall method is a system of movable roof supports that hold up the roof over the immediate work area along the entire length of the face. As the mining machine is advanced into the face the supports are advanced with it, allowing the roof behind the supports to cave in.

Until the mid-1970's longwall mining found little favor in the U.S. It had been tried in the Illinois coal fields as early as 1962, but a series of failures discouraged others from adopting it. In 1975 fewer than 60 faces were in operation. In that year, however, a few mines in Illinois, the site of earlier failures, installed a new system of hydraulically operated shield-type roof supports that had been developed in Europe. By 1979, 91 longwall faces were in operation. At last count there were 105 active longwall faces, with equipment for 21 more on order. In 1981 longwall mines accounted for 18 million tons of coal, or 6.2 percent of underground production. According to some estimates, about 200 longwall faces will be operating in the U.S. in 1985, yielding as much as 12 percent of all underground tonnage.

High productivity is possible in longwall mining. The Sunnyside Mine of the Kaiser Steel Corporation at Sunnyside, Utah, regularly extracts 2,900 tons of coal per day with an 11-man crew from a face 550 feet long, well above the average U.S. productivity. The coal seam, which is nine feet thick, lies 1,500 feet below the surface. In one 24-hour period earlier this year a series of crews working one face produced 20,384 tons of coal, a world record.

The question arises of why the apparently more productive longwall technology has not spread faster in the U.S. One reason is that American mines are generally shallower than those in Britain, West Germany and other countries that are fully committed to the longwall method. Room-and-pillar mining is therefore still economical in most U.S.

CONTINUOUS MINING MACHINES, which first appeared in 1948, also operate in a room-and-pillar mine layout. The machines fracture coal from the face of the seam and load it in a single step. Modern continuous miners are built in several sizes to operate in seams that range from two to 10 feet in thickness. The most widely adopted machines have a rotating drum studded with cutting bits that dig into the coal face. The drum is driven into the top of the seam and travels downward. Gathering arms push the fallen coal onto a central conveyor that discharges the coal onto a haulage system, such as a shuttle car or an extensible conveyor belt coupled to the rear of the continuous miner. After the mining machine has advanced about 20 feet it is withdrawn and moved to another face so that the freshly exposed roof can be bolted. Continuous mining machines now account for about 65 percent of all coal mined underground.

coal seams. Moreover, American mining engineers and workers have had only limited experience with the longwall system. As in other fields, management has learned that pioneering can be costly. The capital investment needed to open a new longwall face is high: from $10,000 to $14,000 per foot of face length, so that a 500-foot face would cost from $5 million to $7 million. In addition continuous mining machines are still needed for development of the mine. Customary marketing arrangements for coal also have an influence on the choice of methods. In the U.S. the output of a mine is usually sold by contract with the assurance of regular shipments to customers. A mine with a single longwall face runs the risk of defaulting on contracts if a mechanical breakdown interrupts operations. To ensure continuity of production the operator must open multiple faces, which raises the capital cost still further.

Because longwall mining is likely to assume increasing importance in the U.S., the experience in the U.K. with longwall mining is pertinent and somewhat sobering. It suggests that the gains from mechanization can be canceled by geological conditions and other factors that are difficult to control. Between 1971 and 1980 the number of operating longwall faces in the U.K. was reduced more than 20 percent, from 840 to 649, reflecting the closure of marginally productive mines. A recent analysis by Brian Lord, chairman of Lord International Mining Associates, shows that in spite of the concentration of output in the better mines, there was no overall improvement in the productivity of longwall mines between 1971 and 1979. It turns out there was a productivity gain of 16 percent at the coal face itself, but it was offset by a decrease in productivity "elsewhere underground." Overall productivity remained constant at 2.24 tons per man-shift.

When Lord classified the longwall mines according to their degree of mechanization, he found that the most mechanized mines were slightly less efficient than the least mechanized ones over the most recent five-year period, 1975 through 1979. The measure of performance was the distance the coal face was advanced in centimeters per man-shift. The least mechanized mines nosed out the most mechanized ones 87 centimeters to 86 centimeters. The best performance was that of mines with intermediate levels of mechanization: 91 centimeters per man-shift.

These surprising results prompted Lord to question whether the British coal industry should pursue the goal of a fully mechanized coal face. He speculates that the industry is already overmechanized and that a simplification might prove beneficial. Between 1946 and 1980 the nationalized British coal industry made a massive effort to modernize its mines. The technological achievement was impressive: the number of miners was reduced by two-thirds, from 711,000 in 1947 to 237,000 in 1980, while output declined only 30 percent, from 187 million long tons in 1947 to 130 million in 1980. Thus production per man-year slightly more than doubled. The economic benefits, however, are questionable: in 1980 the return on average capital invested was exactly zero.

With respect to the mechanization of mines in West Germany similar conclusions have been reached by Gunter B. Fettweis, director of the Institute for Mining Technology of the University of Mining and Metallurgy in Leoben, Austria. He has found that mechanization brings substantial savings initially, when the geological conditions are favorable. As the coal becomes harder to extract, however, the costs in a mechanized mine rise sharply and soon exceed the costs in less mechanized mines where the geological difficulty is comparable. The explanation, Fettweis writes, "is that machines up to now have almost always proved less able to adapt to difficult or changing conditions than people working manually. Under very unfavorable conditions [men and machines] fail together."

American coal operators are beginning to have similar reservations about the benefits of extreme mechanization, according to Paul C. Merritt and David

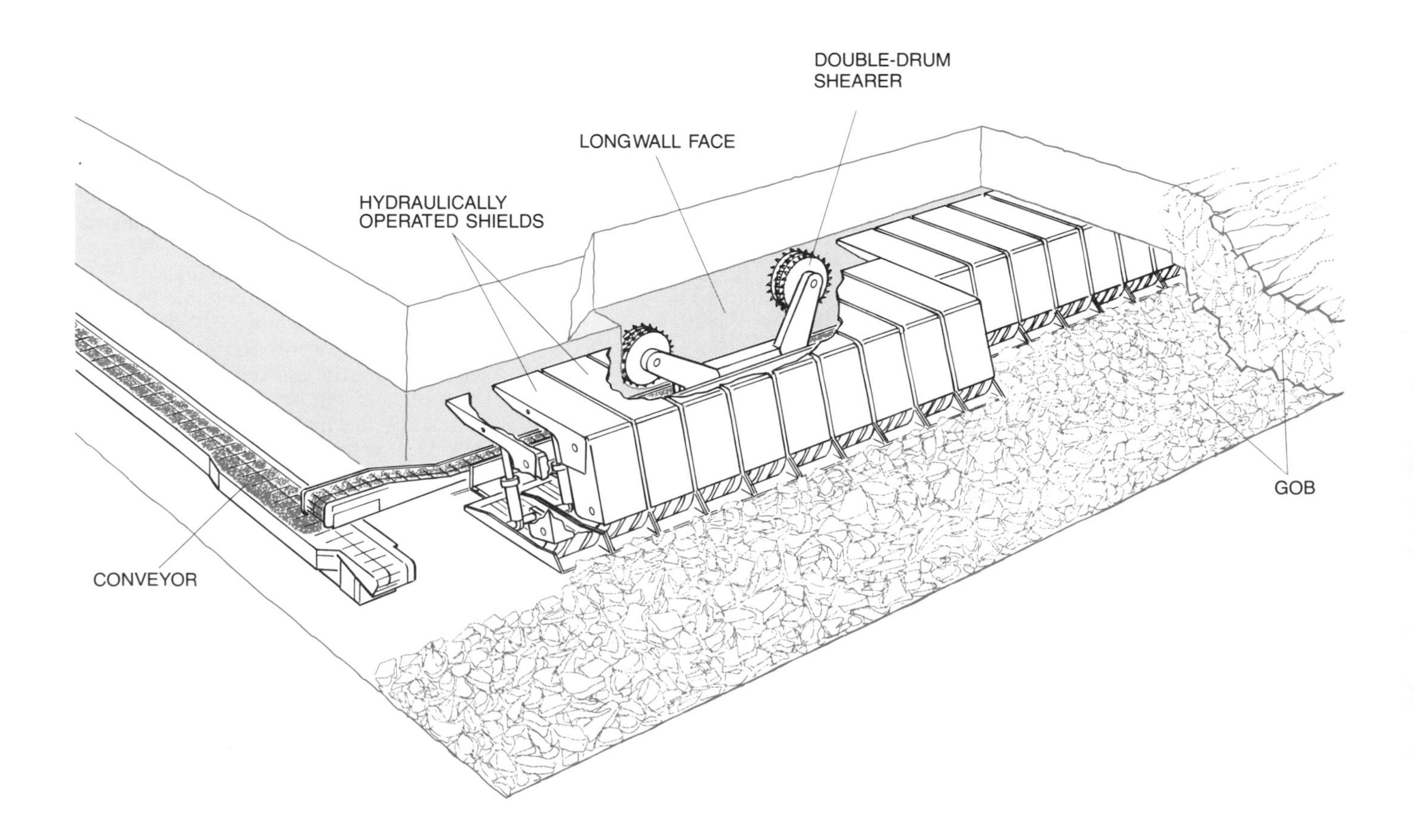

Brezovec of *Coal Age*. Some operators contend that mining is becoming too sophisticated and that the functions given over to machines increase the probability of shutdowns. Individual machines, in many cases those meant to make the job easier, often prove unreliable under actual mining conditions.

The difficult problem of trying to raise mining productivity above levels that are already high suggests to some observers that the roles of private industry and government should be redefined. In some American industries the Federal Government takes a direct hand in research and development; for example, the National Aeronautics and Space Administration contributes to the design of aircraft and the Department of Agriculture contributes to the improvement of crops and animal husbandry. In mining, however, the Government (through the Bureau of Mines) has concentrated more on efforts to improve the health and safety of miners than on efforts to increase productivity. Research on productivity (assumed by the Department of Energy in 1977) has been hampered somewhat by the diversity of conditions encountered in mining. One suggestion is that private industry should concentrate on simplifying and increasing the reliability of existing equipment, whereas government should attend to more speculative areas of technology, including efforts to ensure that greater mechanization will pay off in improved productivity and safety.

One example of a promising idea that proved to be ahead of its time was the "push-button miner." In many areas, and particularly in Appalachia, strip mining can reach only part of a coal seam; operations must be suspended where the ratio of overburden to coal mined exceeds an acceptable value. Additional coal can be recovered by driving large augers into the hillside. In the 1950's and 1960's auger mines accounted for some 10 to 20 million tons of coal per year, or from 2 to 3 percent of total production. The productivity of auger mines was as high as 45 tons per man-day, about 10 tons higher than the peak productivity of strip mining.

In the early 1960's one company tried to develop an automatically controlled coal auger: the push-button miner. In field tests the performance of the machine was impressive: coal could be extracted at a rate of almost 500,000 tons per year. Under actual mining conditions, however, the push-button miner proved unreliable and could not exceed the performance of conventional augers. Only two of the machines were built, and development was abandoned.

In 1981 a machine with features similar to those of the push-button miner was introduced into the U.S. by Rhine-Schelde-Verolme Machine Fabriken en Scheepswerken N.V., a shipbuilding company in the Netherlands. The machine, known as the Thin-Seam Miner, is designed to mine hillside coal seams from 24 to 63 inches high. The cutting head tracks the coal seam with the aid of sensors that distinguish the differing levels of natural radioactivity in the coal and in the surrounding layers of shale. The reach of the cutting head is increased by the insertion of 20-foot extension beams that allow coal to be mined to a depth of 220 feet.

The manufacturer states that the miner can recover up to 85 percent of the coal within its reach. With a crew of four the machine is said to be capable of mining 420 tons in an eight-hour shift, or 105 tons per man-shift. Several Thin-Seam Miners have been operating in the U.S. since last fall. They are valued at $2.5 million each and are leased rather than sold. In return for royalties on the coal mined the distributor provides a complete service, including installation, operating crew and maintenance.

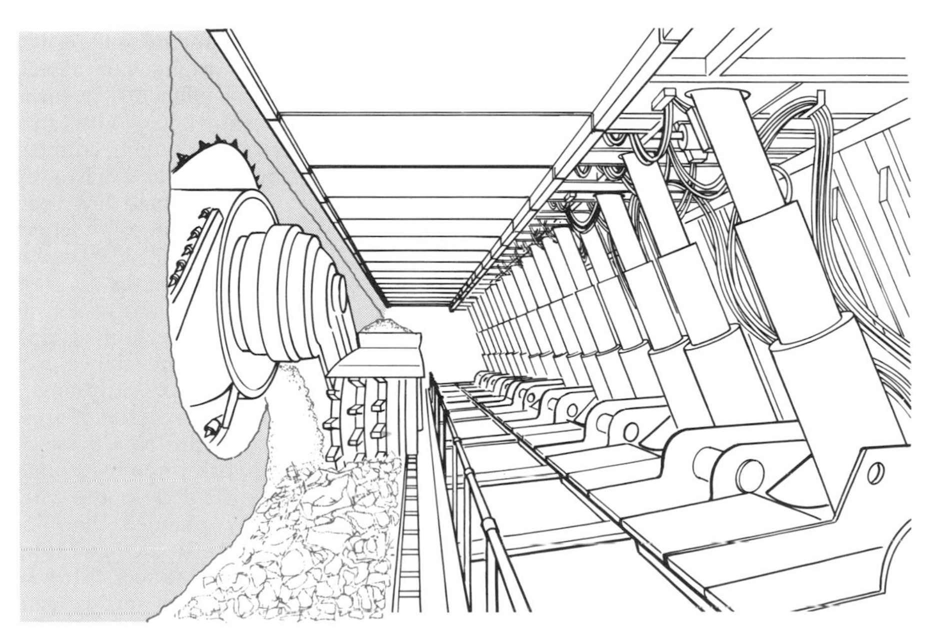

LONGWALL MINING, which has been common in British and European mines for many years, exploits a continuous mining machine that either planes or shears coal from one face of a block 500 feet wide and up to a mile long. The machine shown is a double-drum shearer. The cutting machine makes continuous passes across the entire face. The mine layout no longer follows a room-and-pillar pattern (except for entryways surrounding the longwall panels). Instead the roof adjacent to the longwall is supported by hydraulic props that move forward as the face is advanced. The roof behind the props is allowed to collapse, leaving rubble called gob. When a panel has been mined out, the cutter and supports are moved to the next panel.

In conventional surface mining the basic sequence of operations—stripping away the overburden and mining the exposed coal seam—has not changed from the beginning. As equipment has grown larger and more powerful it has become economic to mine seams as much as 200 feet below the surface with draglines and excavators whose buckets can hold up to 180 cubic yards of earth and rock. To loosen the overburden drills now sink holes more than 10 inches in diameter to a depth of 75 feet to accept explosives fed in bulk from large trucks. The coal seam is fractured in the same way. The coal itself is scooped up by shovels with a capacity of up to 22 cubic yards and is loaded into trucks that hold up to 170 tons.

Increases in the size of surface-mine equipment are now slowing. Machine improvements are directed instead toward efficiency, reliability and ease of maintenance. Major subassemblies are being made in separate modules to facilitate transport, initial installation and subsequent repair.

Diagnostic tools, equipment monitoring and operator aids have become increasingly important as the cost of fuel and the cost of having machinery out of service for maintenance or repair have mounted. Digital computers have entered mining as they have every other industry. A dragline simulator, for example, has been programmed to teach a new operator the "feel" of the device before he climbs into the cab of the actual machine. Once the operator is at work his skills are improved by a feedback system. An on-board computer provides a visual display of performance and retains a record of machine operation and condition, thereby helping to optimize output and decrease maintenance.

Although most equipment used for

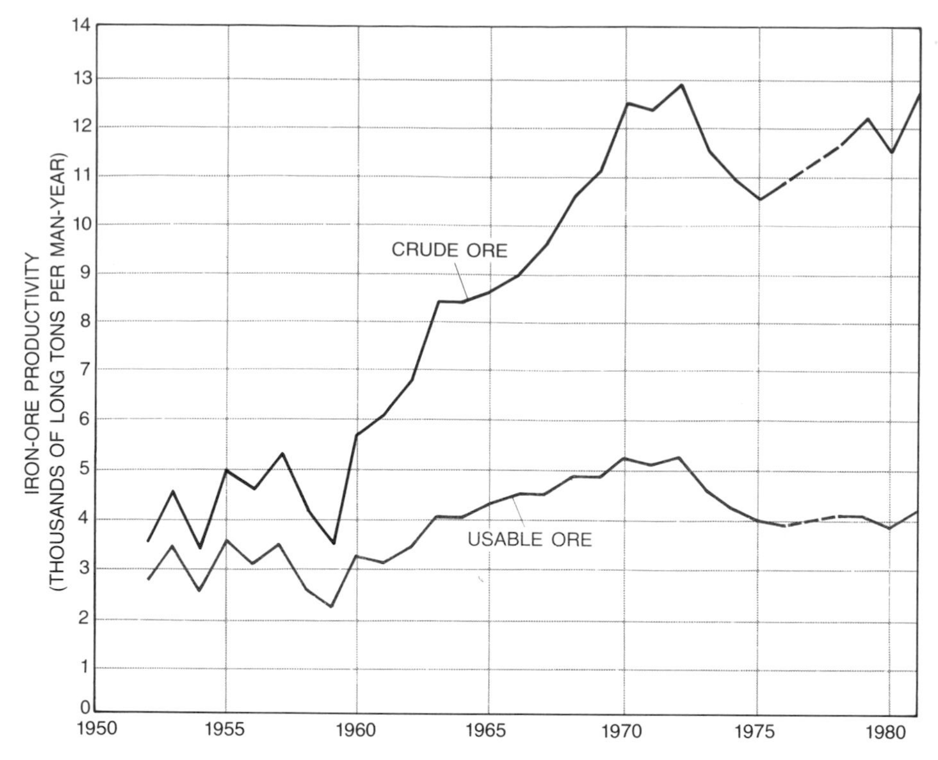

PRODUCTIVITY IN IRON-ORE MINING between 1952 and 1981 increased at an average rate of 4.5 percent per year calculated on the basis of crude ore mined (*black curve*) but at the much lower rate of 1.5 percent when measured on the basis of usable ore (*color*). Most domestic iron ore comes from surface mines. The 1981 output of 4,180 long tons (4,680 short tons) per man-year of usable ore can be compared with the productivity of 6,700 short tons per man-year in surface coal mining, where productivity has increased at a rate of 2 percent per year.

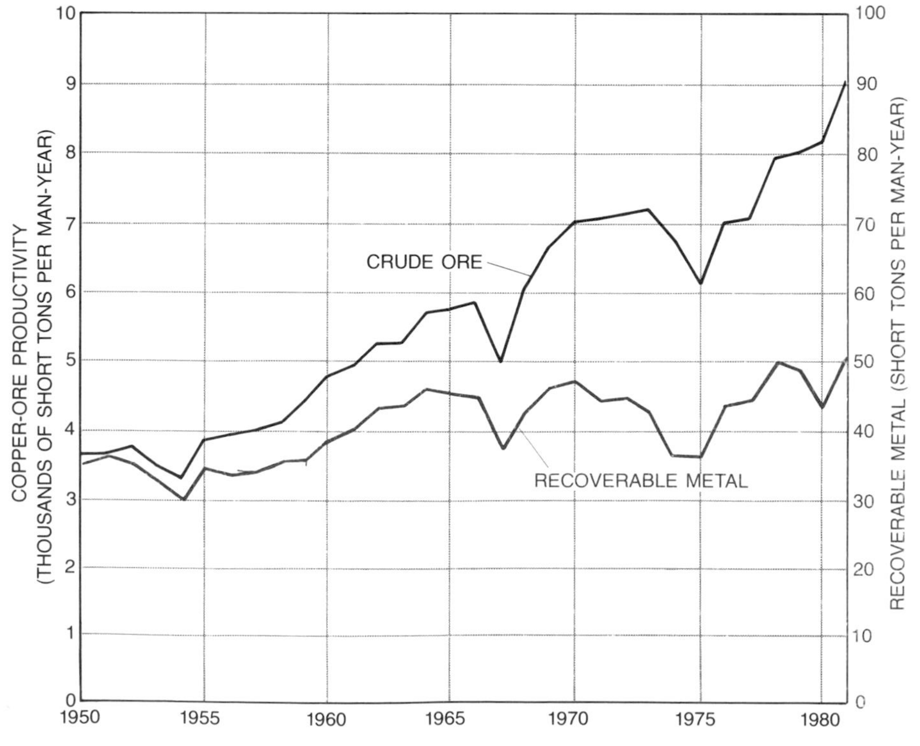

PRODUCTIVITY IN COPPER-ORE MINING between 1950 and 1981 increased at a rate of 3 percent per year on a crude-ore basis (*black curve*) and at a rate of 1.2 percent per year for recoverable metal (*color*). About 85 percent of the copper-ore tonnage is extracted from open-pit mines. In 1950 each ton of copper ore mined contained an average of about 19 pounds of copper. By 1981 the average metal content had dropped to about 11 pounds per ton of ore.

surface coal mining could serve for general excavation and earth-moving work, one machine is highly specialized. It is the bucket-wheel excavator, which holds great promise for efficient extraction of Western coals and Texas lignite that are somewhat lower in energy content than those customarily mined. These huge machines, built in West Germany by O & K Orenstein & Koppel Aktiengesellschaft, have bucket wheels from 7.8 to 15 meters in diameter. They can rapidly remove a large volume of overburden and deposit it several thousand feet away by means of conveyors without the need for truck haulage or other rehandling of the material.

The bucket-wheel excavator makes possible the recovery of seams that lie more than 300 feet below the surface, well below the depths now considered feasible for stripping. In an eight-hour shift the machine can remove as much as 24,000 cubic meters of overburden, the equivalent of a football field excavated to a depth of 15 feet. Although the machines cost from $1.5 to $4 million, depending on their size, they should prove economical for mines that produce two million tons per year or more, particularly if the customer is a nearby electric-power plant. Under such conditions competitively priced electric power could be generated from coals and lignite of low energy value.

Strip-mine operators are required by Federal and state laws to return mined-out areas to their approximate original contour and to make them fit for their former use. Power requirements for leveling the loose mine wastes are much lower than they are for dislodging the overburden from its original compacted state. The job is done efficiently by large bulldozers. In recent trials two bulldozers have been yoked in tandem, coupled to a single 40-foot blade, and driven by one operator. Once the land has been graded it is reseeded. Power mulchers chop hay and spray it on hillsides to prevent erosion. In many cases aircraft broadcast seed, fertilizer and mulch. It is not unknown for the reclaimed land to have a higher value and higher productivity than it had before stripping.

The surface-mine work force must have skills very different from those of the underground work force. Although functions directly related to the coal itself are generally planned by mining engineers, many of the operators of draglines, shovels, graders, trucks and bulldozers working in surface coal mines have had no mining experience. They have worked instead in other large earth-moving projects.

Over the years the benefits of mine mechanization in raising both the productivity of workers and their income have been offset in part by new hazards introduced by changes in technology. Beginning in 1870, when records were first kept, the number of workers killed in coal-mine accidents in the U.S. initially dropped and then rose. The 1880 fatality rate of 2.2 deaths per 1,000 workers was exceeded in 52 of the next 53 years. Only since 1948 has the fatality rate stayed below two per 1,000, with the exception of 1968, when it again reached 2.2. Even so, the overall trend since 1907 (the worst year of the century, when at least 3,024 miners died in underground-mine accidents) has been downward. Since 1970 the number of coal miners killed has averaged 140 per year, a significant drop.

The most serious hazard introduced by mechanization was fine coal dust, which became a problem when the pick was replaced by power machinery for undercutting the coal face. The hazard is increased when methane gas trapped in the coal is liberated as the coal is fractured. When methane is mixed with air in concentrations of from 5 to 15 percent, it is highly explosive. The explosion of a small amount of methane can trigger a much larger explosion of coal dust, or even a series of explosions that might be propagated throughout the mine. Very likely it was such a series of explosions that killed 358 miners in a West Virginia mine in December, 1907, the record number for a single coal-mine disaster.

The first undercutting machines increased the rate of coal fracture and with it the rate of methane liberation. This led to the need for improved ventilation. The problem became worse still with the adoption of continuous mining machines, because the coal not only was cut faster but also was ground into finer particles and more dust. Mining machines are now equipped with methane sensors that shut the units down before the concentration reaches a dangerous level. The cutting is also done under a continuous water spray to hold down dust. The main precaution against dust accumulation, however, is the application of an inert layer of limestone over the walls and roof as mining proceeds.

Much of the mine-safety legislation in the U.S. has been stimulated by mine disasters. Over most of its history the Bureau of Mines has been concerned with the prevention of mine fires and explosions and with improving mine rescue operations. The Federal Coal Mine Health and Safety Act of 1969, and amendments to it enacted in 1977, gave the bureau responsibility for all aspects of mine safety and miners' health. The legislation was precipitated in part by the efforts of the United Mine Workers of America to gain compensation for miners who had developed black-lung disease, a respiratory disease caused by prolonged inhalation of coal dust. Another factor was a series of mine accidents that caused 311 fatalities in 1968,

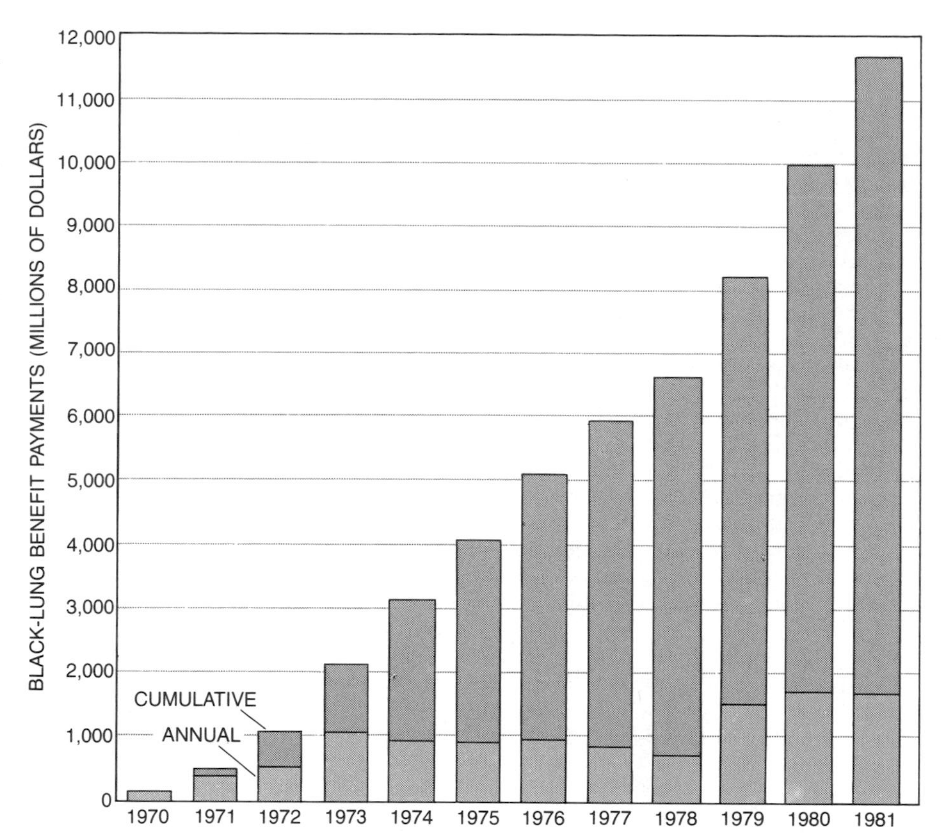

BENEFITS TO RETIRED MINERS suffering from black-lung disease are financed in part by a levy of 50 cents per ton on surface-mined coal and $1 per ton on coal mined underground. Last year benefits of $600 million were paid from this fund to about 90,000 miners, dependents and survivors. An additional $1.1 billion was paid in the form of Social Security benefits.

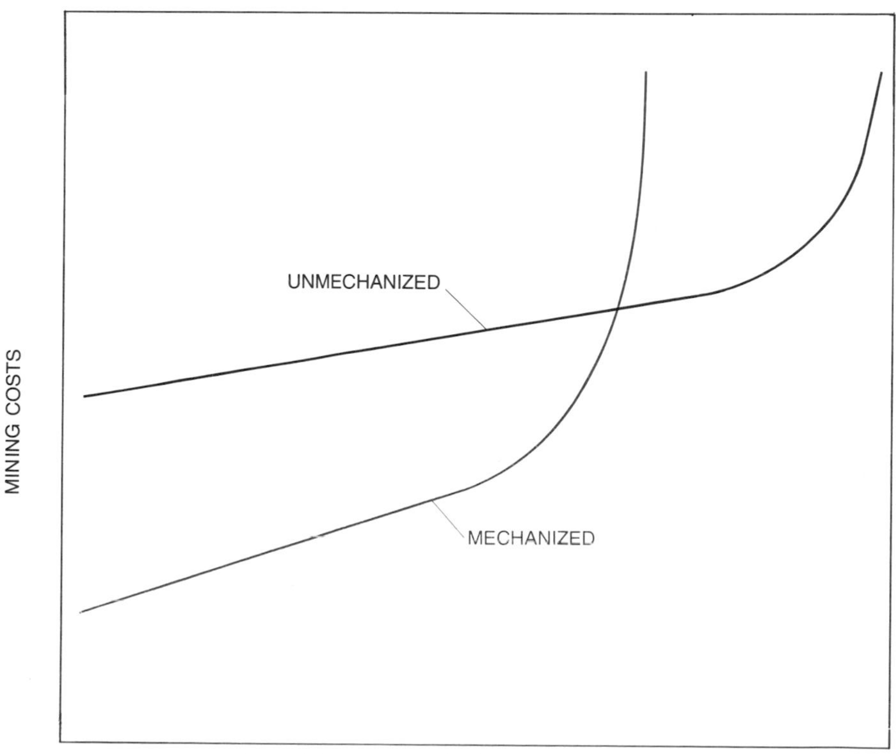

STUDY OF MECHANIZATION IN GERMAN COAL MINES has shown that the benefits of mechanization diminish as mining is extended to coal deposits that are less accessible and of lower quality. Eventually the costs per unit of coal extracted in a mechanized mine exceed those in an unmechanized mine exploiting comparable deposits. The diagram is from a report by Gunter B. Fettweis of the University of Mining and Metallurgy in Leoben, Austria.

including an explosion in Farmington, W.Va., that killed 78 miners.

Respirable dust in underground mines continues to be a major health problem, and so does noise. It is well documented that exposure to high noise levels over a long period leads to permanent hearing impairment. The aggregate cost to the miner, his family and society cannot readily be calculated but is certainly substantial. Coal mining, of course, is not the only industry where sustained noise damages the hearing of its workers. Since 1970 the coal industry has made considerable progress in reducing the noise of its machines, above the ground as well as below. So far miners whose hearing has been impaired have not been entitled to compensation in the U.S. In Australia, on the other hand, the number of claims for occupational deafness has risen steeply, reaching 600 in 1979 and 1980.

The U.S. began paying benefits to retired miners with black-lung disease in 1970. In the first five years the cumulative payments amounted to $3.1 billion. In the next five years another $5.1 billion was paid out. To date the program has cost $11.7 billion.

The 1969 Health and Safety Act specifies that the respirable dust in coal mines must not exceed two milligrams per cubic meter of air. Room-and-pillar mines, including those with continuous mining machines, have been able to satisfy the regulation. In longwall mines, however, at any one time about half of the operations are having difficulty complying. The Government and industry have joined forces to solve the problem. In 1970 about 15 percent of underground coal miners exhibited some degree of black-lung disease. The incidence is now down to about 6 percent.

Changes in the technology of mining through the remainder of the century will be evolutionary, as they were in the past, rather than revolutionary. A period of at least 10 to 20 years is needed for a new mining technology to replace an older one. The reasons are the same as those that could be cited in any other industry: the high capital cost of new equipment, the inclination of operators to stick with proved methods and uncertainty about the performance of new devices. A special obstacle to innovation in underground coal mining is that a change in a subsystem, such as haulage, may require a change in the entire mining system. Nevertheless, longwall mining, which requires the most radical change of all, is expected to grow, depending in part on how successfully it can be adapted to thin seams in the East and thick ones in the West.

The future of surface mining will be influenced by changed perceptions of the economies of scale. At some point the operator finds that too much capital has been invested in equipment. The trend to ever larger surface equipment can therefore be expected to slow. Ultimately as operators of surface mines attempt to reach deeper seams they will find their costs approaching the costs of underground mining. Then the half-century decline in underground mining will begin to reverse.

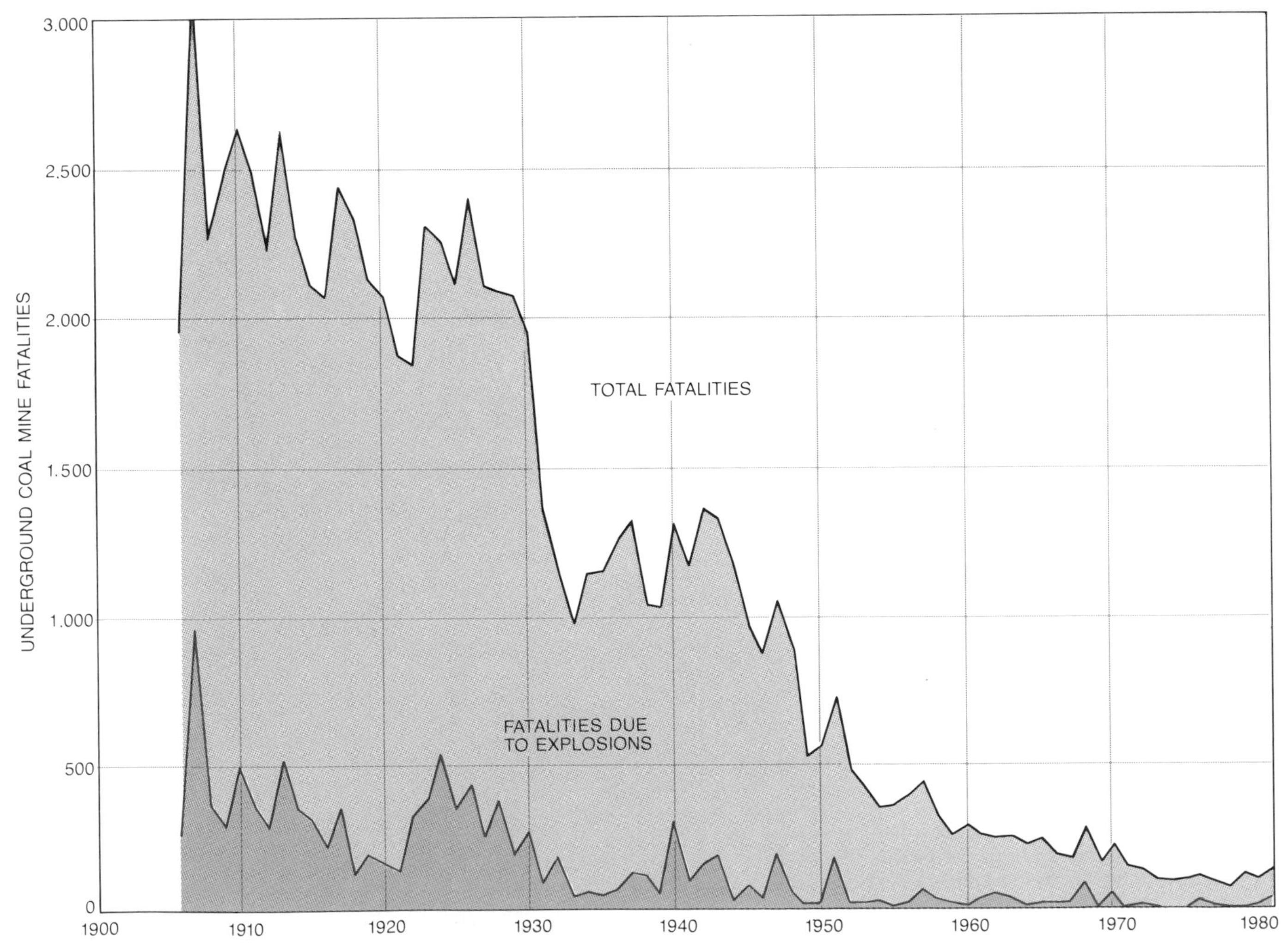

FATALITY RATE IN UNDERGROUND COAL MINES has declined sharply in the past 50 years with improved safety measures. Most fatal accidents in underground mines are caused by roof falls and cave-ins, which regularly took from 1,000 to 1,200 lives per year until 1931, but explosions are dreaded because of their potential for killing many miners in a single accident. Explosions can result when methane released from the fractured coal reaches a dangerous concentration. A methane explosion can trigger explosions of coal dust.

4

THE MECHANIZATION OF DESIGN AND MANUFACTURING

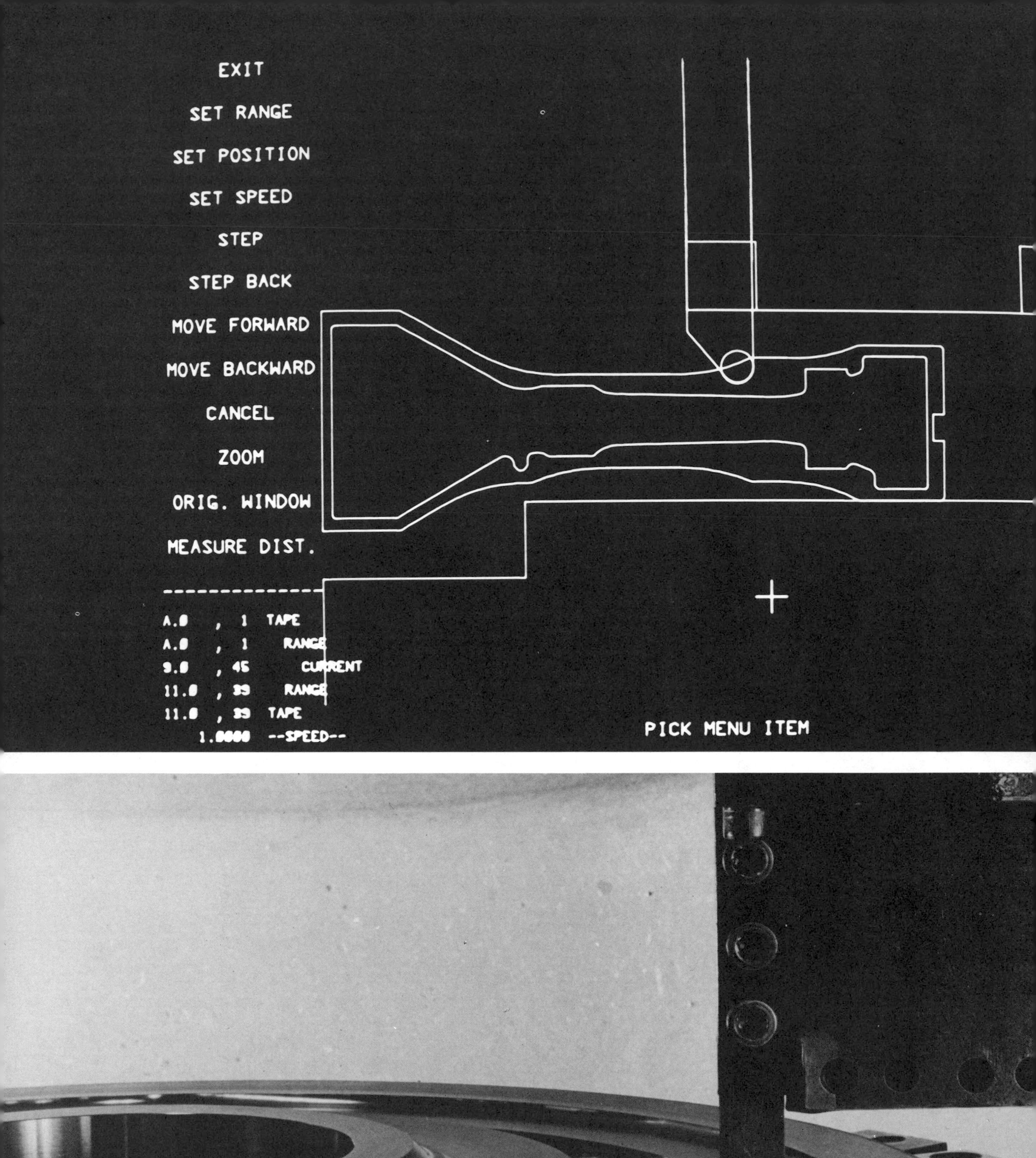

EXIT
SET RANGE
SET POSITION
SET SPEED
STEP
STEP BACK
MOVE FORWARD
MOVE BACKWARD
CANCEL
ZOOM
ORIG. WINDOW
MEASURE DIST.
A.0 , 1 TAPE
A.0 , 1 RANGE
9.0 , 45 CURRENT
11.0 , 39 RANGE
11.0 , 39 TAPE
1.0000 --SPEED--
PICK MENU ITEM

The Mechanization of Design and Manufacturing

by THOMAS G. GUNN

Mechanization on the factory floor continues, but greater changes are being introduced by new technology for the design of products and for planning, managing and coordinating their manufacture

The factory is the workplace where mechanization might seem to be most deeply embedded. Indeed, the origin of the modern factory can be traced to the introduction of water-powered and steam-powered machinery in the 19th century, most notably in the textile industry. Today the typical factory relies on a great variety of machines, and factory work is widely supposed to consist largely of tending machines. Newer developments in the relations between men and machines also tend to be viewed primarily in the context of manufacturing, and the social consequences of mechanization have been considered chiefly through their effects on production workers. There is an irony here: it turns out that manufacturing is one of the most difficult sectors of the economy in which to realize the full potential of the available technology.

The opportunities for mechanization in the factory are widely misunderstood. The emphasis has been almost exclusively on the production process itself, and complete mechanization has come to be symbolized by the industrial robot, a machine designed to replace the production worker one for one. Actually the direct work of making or assembling a product is not where mechanization is now likely to have the greatest effect. Direct labor accounts for only 10 to 25 percent of the cost of manufacturing, and workers engaged in such tasks make up only two-thirds of the total manufacturing work force. The major challenge now, and the major opportunity for improved productivity, is in organizing, scheduling and managing the total manufacturing enterprise, from product design to fabrication, distribution and field service. The complexity of the modern factory is daunting: in some plants thousands of parts must be kept in stock for hundreds of products. Indeed, the complexity of the operations has sometimes led to a situation resembling gridlock on the factory floor: it is not uncommon for a metal part to spend 95 percent of the time required for its manufacture waiting in line for processing.

Thus the productivity of the factory worker depends in large measure on the design of the product and on the way the resources of labor, machines and raw materials are brought together. Without improvement in these functions it is not clear that even a total replacement of blue-collar workers by robots would have much effect on the output of the factory or on the cost of its products. For this reason the most important contribution to the productivity of the factory offered by new data-processing technology is its capacity to link design, management and manufacturing into a network of commonly available information. The social outcome of the linkage may be to alter far more white-collar jobs than blue-collar ones.

In the U.S., companies whose primary business is manufacturing employ some 20 million people, or roughly a fifth of the labor force. The proportion has been declining for the past 40 years, partly because of the shift of the American economy away from the production of goods and toward the production of services and partly because of the technological displacement of workers.

Manufacturing includes a vast array of activities and enterprises. In some industries a product is made in a continuous stream; familiar examples are petroleum refining, papermaking and the manufacture of many chemicals. A distinctive feature of such processes is the ease with which they can be adapted to control by closed-loop feedback systems: changes in the nature of the product can be detected and employed to adjust the input of raw materials or the intermediate steps in manufacturing. Other process industries, such as steelmaking and brewing, generally work with batches of materials.

Here I shall focus on industries of another kind: those that design and manufacture discrete products rather than materials that are processed continuously or in batches. Discrete-products manufacturing is a broad and varied category. It encompasses the fabrication and assembly of automobiles, aircraft, computers and the microelectronic components of computers, furniture, appli-

COMPUTATIONAL LINK between design and manufacturing is illustrated by a system employed by the General Electric Company to control the cutting path of a vertical turret lathe as it shapes a component of an aircraft engine. In the upper photograph two cross sections passing through the radial axis of the disk-shaped part are shown on the screen of a computer-aided-design terminal. The outer cross section shows the shape of the forged part before it is machined; the inner cross section shows the final shape of the part. Above the cross sections the cutting tool of the lathe is represented schematically in one of the positions it assumes while making the cut. The path of the cutting tool is controlled by a computer program; the programmer, who must know the precise geometry of the part in order to program the cutting tool, can retrieve the information from a central data base where it is stored during the design of the part. The path of the tool is then animated on the screen so that the programmer can verify that the cut will proceed as intended and without interference. The lower photograph shows the real cutting tool in a position corresponding to the one simulated on the computer terminal. To make the cut the disk is rotated on a turntable under the cutting tool, which is moved up, down and across the disk according to the pattern determined by the computer program. The disk is made of a titanium alloy and is part of the high-pressure compressor of the CF6-50 turbofan engine. The engine powers a number of commercial and military jet aircraft, including the Boeing 747 and the McDonnell Douglas DC-10; it delivers a thrust of 50,000 pounds.

ances, foods, clothing, packaging, building materials and machine tools. It is in these industries and the many like them that the potential benefits of data-processing technology are most pronounced. Only in the past 20 years has the discrete-products manufacturer had the means to gain feedback from his operations (and hence continuous control over them) by methods analogous to those of the process manufacturer.

An example of a discrete-products factory I shall frequently refer to is the machine shop. There metal parts are made by a sequence of operations, including cutting, boring, milling, grinding and turning on a lathe. The same set of machine tools can serve to make a great variety of parts, from gun barrels to camshafts; what changes is the sequence of operations carried out with each tool and the sequence of tools employed. The very versatility of the tools makes the efficient organization of machine-shop operations difficult. Methods of programming, or numerically controlling, the machine tools themselves were introduced some time ago and have been widely adopted. The emphasis now is on coordinating the operations of various machines and controlling the flow of work through the shop.

Over the years manufacturers have sought to plan, control and carry out complex operations by setting up large bureaucracies with many levels of responsibility. In some American companies there are as many as 14 layers of personnel in the chain of command from the chief executive officer to the lowest-ranking shop-floor worker. The hierarchical structure enables a company's management to direct the overall operations of the company, but the organizational distance between the top and the bottom makes it difficult to keep close track of labor and plant resources, to schedule their deployment and to control job priorities. Moreover, a hierarchy can give rise to institutional barriers between departments and can inhibit the internal flow of information. It can also create a competitive "us against them" attitude among departments that hinders the functioning of the organization as a whole.

The organizational barriers to communication are often supplemented by physical ones. Written and verbally transmitted information is subject to delay, error and misunderstanding as it passes from one person to another. The time required for a memorandum to circulate can make prompt collective action impossible. As a result large manufacturing organizations have responded sluggishly to changing market conditions. Another consequence of impediments to communication is a need to stock large inventories of products, parts and materials at considerable cost in space, insurance and handling.

By the 1950's the burden of paperwork and verbal communication in many manufacturing concerns had become onerous. In order to get work done on time the companies began to employ expediters who operated independently of the organizational hierarchy to push products out the factory door.

Manufacturing companies were introduced to the computer in the early 1960's. The first applications were the recording of routine financial transactions, but gradually computers were applied to other tasks, such as the control of inventory, the scheduling of production and the routing of a part from one process to another on the factory floor. As the applications diversified and various departments of the company adapted the computer to their needs it became apparent that the advantages of computing technology within each department could be multiplied many times if

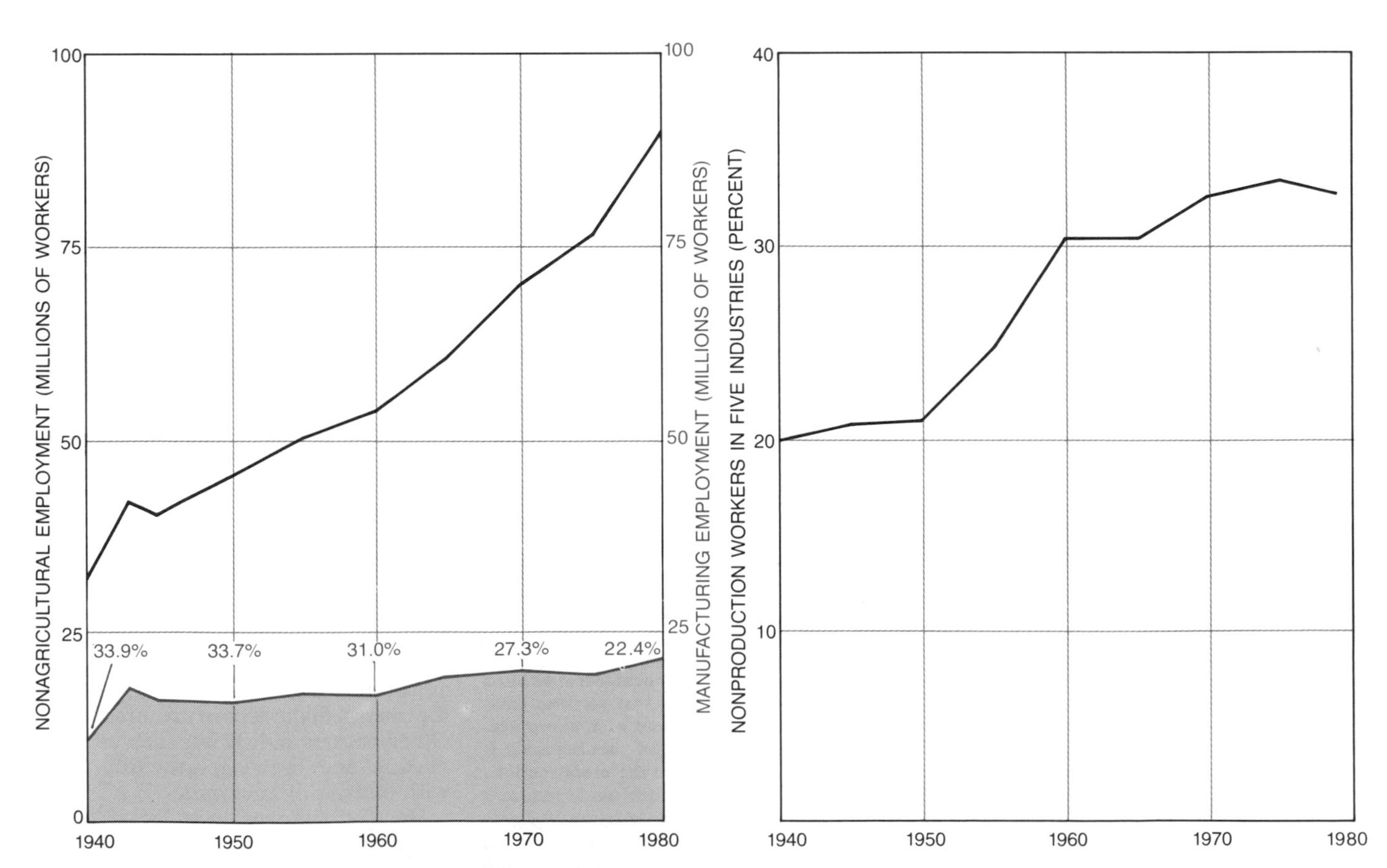

FRACTION OF MANUFACTURING WORKERS in the U.S. engaged in indirect labor (such as planners, expediters, salesmen, managers and the like) has increased even as the manufacturing fraction of the work force has been reduced. There are now about 20 million workers in manufacturing, or 22 percent of the total number of employees on nonagricultural payrolls. In the five most highly mechanized industries, namely fabricated metal products, machinery, electric and electronic equipment, transportation equipment and instruments and related products, workers who supply direct labor currently account for only about two-thirds of the total work force.

certain departments or functions were linked.

The earliest functions of the computer that had a direct bearing on manufacturing operations were not in the manufacturing process itself but in the design of products. In the mid-1960's engineers at the General Motors Corporation began working with programming specialists at the International Business Machines Corporation to develop a system for computer-aided design. The system was originally envisioned only as a sophisticated drafting tool. The design engineer would employ a keyboard to specify certain numerical data about the part, but he could touch a light-sensitive stylus directly to the screen of a cathode-ray tube in order to "draw," or enter geometric data into the computer. Although the motion of the stylus on the screen might correspond only roughly to the shape of the part, the computer was programmed to combine the numerical and the geometric data so that the designer's sketch could be transformed rapidly into a precise engineering drawing. Because the drawing was stored in the memory of the computer it could be recalled at any time.

The information specifying the geometry of a part is also needed to determine how a cutting machine, such as a lathe, must be operated to shape the part. (The specification of a cutting path must also take into account the capacity of the cutting machine, the material from which the part is made, the shape of the cutting tool, the speed and depth of the cut and other variables.) Traditionally the machinist set up his machine according to drawings supplied by the designer; when numerically controlled machine tools were introduced, the programmer who prepared the sequence of instructions still obtained geometric information from drawings. Designers and programmers soon recognized, however, that the programmer could get the part geometry directly from the data base after it was entered into a computer by the designer, and engineering drawings could be eliminated. Indeed, in many circumstances the programming of machine-tool operations is so routine that little human intervention is necessary once the part geometry is known.

The need for similar information in designing a part and in programming a machine tool to make it illustrates another important contribution that computerized information links can make to manufacturing productivity. Before the advent of computers information pertaining to a product was often scattered in various departments of a company. The engineering drawings, for example, carried certain descriptive information about a product as well as its geometry, but details about how the

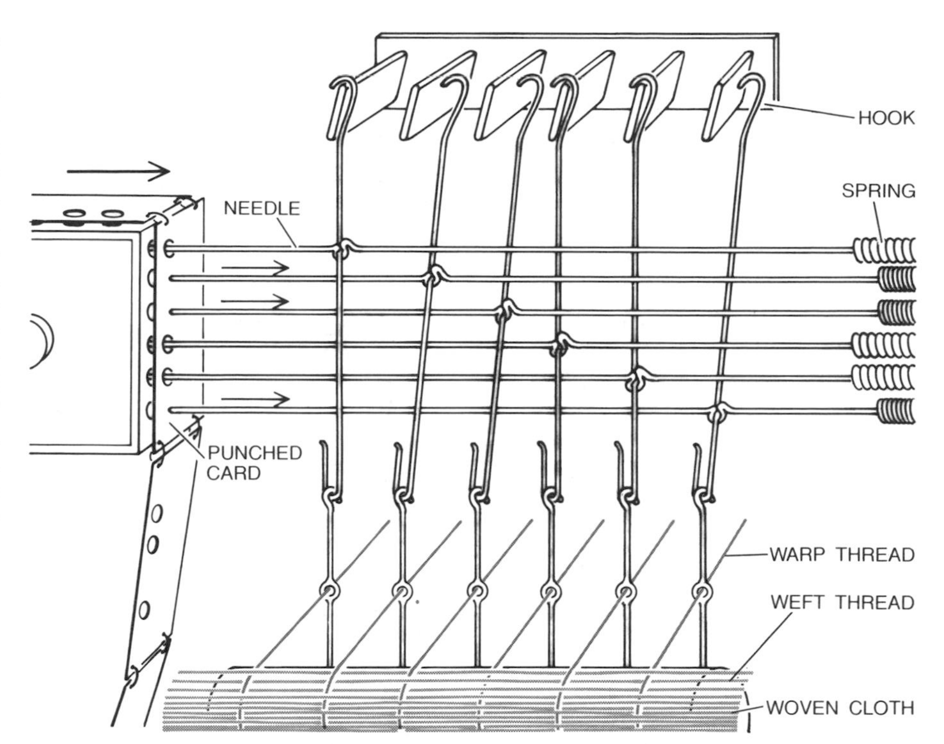

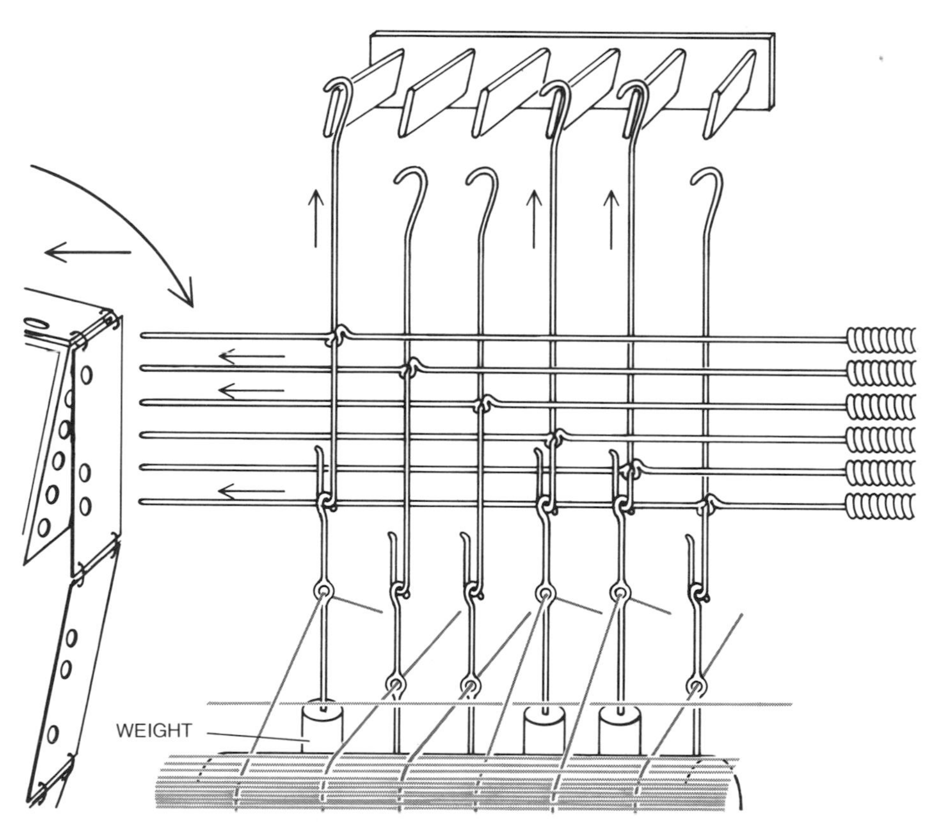

JACQUARD LOOM was the first practical industrial machine to be controlled by a stored program of operations; the program was encoded in an array of holes punched in a series of cards. The Jacquard loom was the forerunner of the numerically controlled machine tool, which carries out instructions encoded as holes punched in a paper tape. On the loom the pattern of the weave is ingeniously made to correspond to the pattern of the holes on the cards. A row of needles mechanically connected to the warp, or lengthwise, threads probes each row of holes. If a needle engages a hole, the corresponding thread is raised during the next operating cycle and so appears on the upper surface of the fabric. If the needle does not engage a hole, the thread remains in place and passes to the underside of the fabric. The programming cards enable the machine to execute a complex sequence of motions automatically, and they allow the sequence to be altered conveniently by changing the pattern of holes. The device was invented for the silk industry by the French weaver Joseph-Marie Charles Jacquard in 1801.

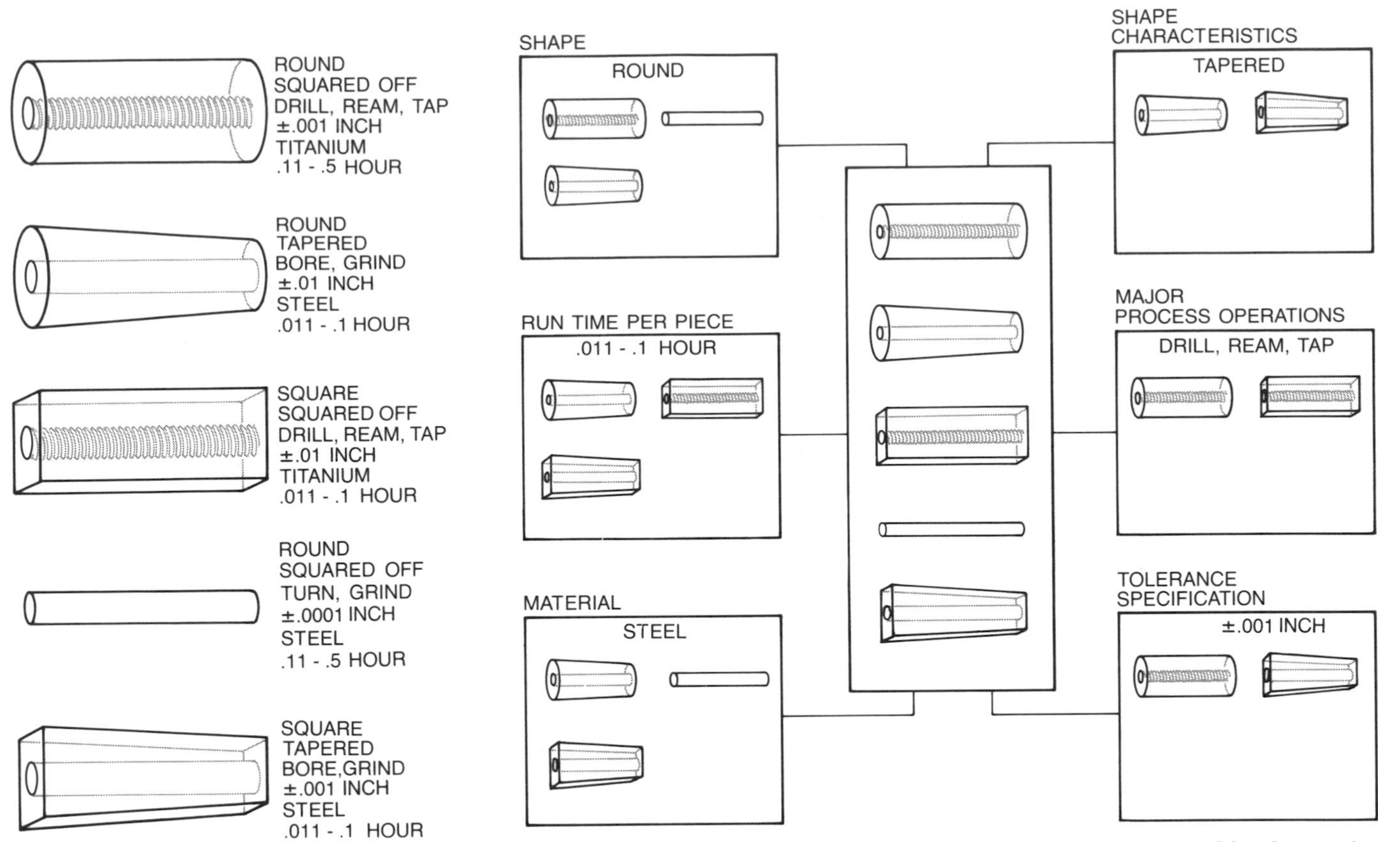

BENEFITS OF MECHANIZATION in the factory can result from more efficient management of information as well as from the automatic control of production machinery. Group technology is an electronic filing system whereby the names of all the parts made by a manufacturing company are stored with a list of descriptive characteristics of each part. The parts are cross-referenced by characteristics such as shape, material and processing operations, and so a list of all the parts that have given characteristics can be generated. Group technology can eliminate the wasteful design of a new part whenever it identifies a part already made that will serve the same purpose.

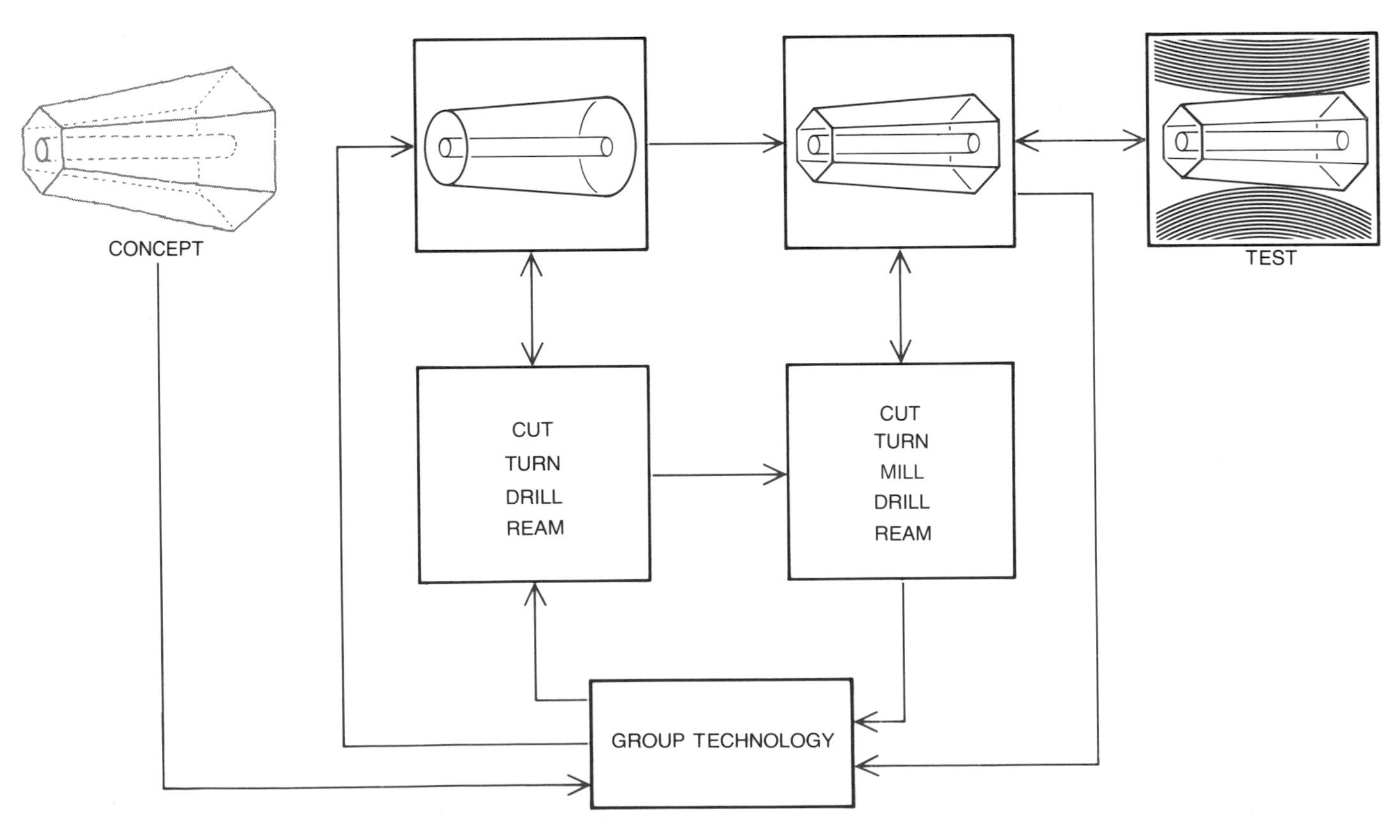

SIMPLIFIED DESIGN and process planning are made possible by group technology. If a new part is needed, the designer can search the inventory of old parts stored in the electronic group file according to the characteristics specified for the new part. For example, if the designer wants a part that is tapered, squared off at the ends, made of steel and bored through along the radial axis, the file might retrieve a bored, truncated conical part already being manufactured, together with the processing steps in its fabrication. The designer can then modify the old design and process plan to create the new part without starting from scratch. The new design can be tested with the help of a computer program to meet engineering specifications; the design and process plan are then stored in the group-technology file.

product was to be manufactured, what machines were to be used and when the product was scheduled for processing on a particular machine were kept only by the department responsible for that aspect of production. Much of the scattered information was redundant, and its distribution created difficulties in keeping it accurate and up-to-date.

My colleagues and I have identified six functional areas that are now being linked to manage the flow of information throughout a factory. The areas are design, the storage and retrieval of information about the parts being manufactured, the management and control of available resources (such as labor, machines and materials) according to changing demands, the handling of materials, the control of machine tools and other single-purpose machinery and the control of robots. By linking the six areas one can achieve what Joseph Harrington, Jr., of Arthur D. Little, Inc., has called computer-integrated manufacturing. It is important to realize, however, that data-processing technology must be fairly well developed in each of the areas before the benefits of linking them become significant.

Perhaps the most remarkable instance of growth in productivity as a result of information technology is in the design of parts and production processes. Computer-aided-design programs can carry out geometric transformations so fast that the designer is no longer limited to the top, side and front views of a part that were characteristic of manually prepared drawings. He can observe the rotation of the part about any axis on the screen, "zoom" in close to see details or take up a distant point of view to visualize the object as a whole. Any cross section of the part can be displayed. If the part is to be mated with other parts during assembly, the designer can move the parts about on his screen to check for fit. Hence many prototypes and engineering models can be eliminated.

The image displayed on the screen can be stored permanently in the data on a magnetic tape or disk. If a copy on paper is needed, it can be generated quickly with a plotting device driven by the computer. Because the design is simple to alter in electronic form it can be changed as many times as necessary without the major effort of redrawing. The design is accessible to everyone who must work with it as soon as it is electronically filed, so that manufacturing functions such as the planning and scheduling of production can be started earlier. Because alterations are made only to the centrally filed design there is less chance that someone will work with an outdated version.

The ready accessibility of the design throughout the company tends to break down the institutional barriers between

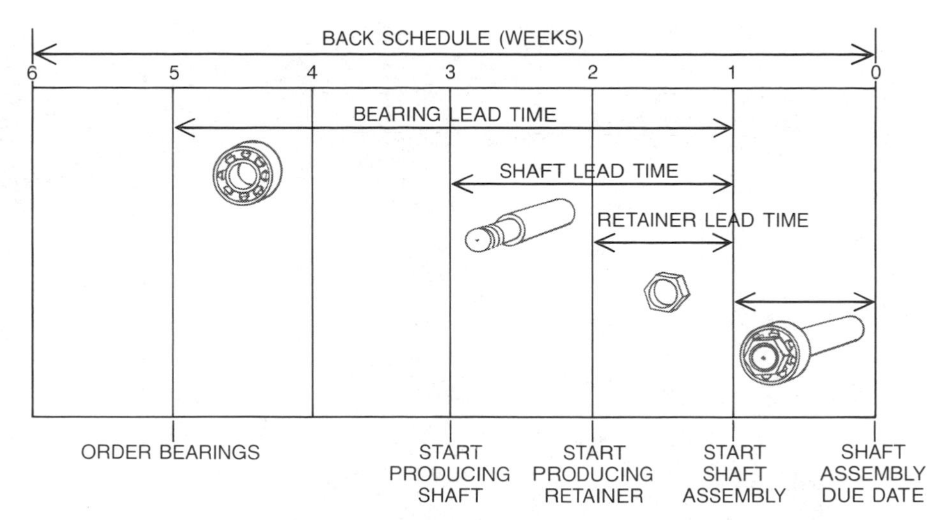

BACK-SCHEDULING from the date a finished product is needed must be done in order to determine when the component parts must be manufactured or bought from suppliers. If the component is ready on the date it is needed and not before, the expenses associated with maintaining an inventory of parts (such as insurance, warehousing and the cost of capital) can largely be eliminated. To reduce inventory without causing delays, however, the back-scheduling must be thorough and accurate. The illustration shows how back-scheduling might be accomplished for a part made up of three components. Accurate back-scheduling depends on careful analysis of the lead time and manufacturing time for each part and on an accurate inventory.

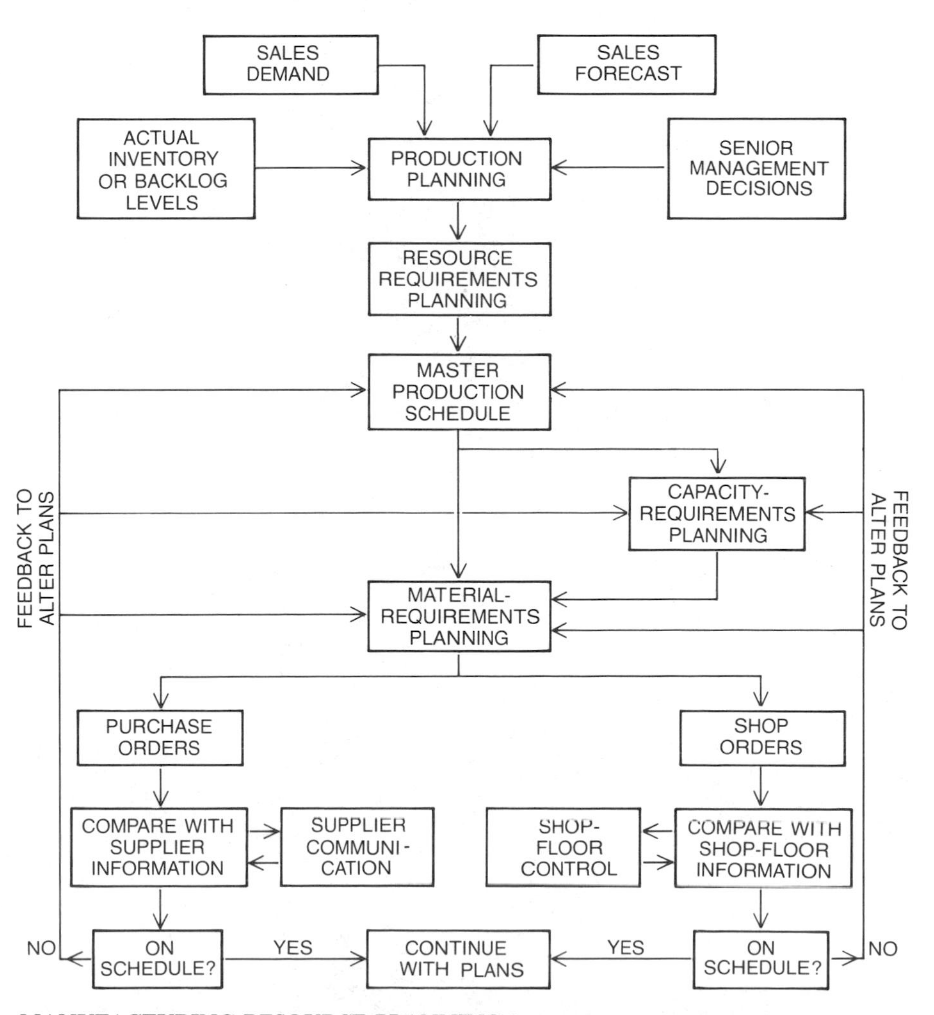

MANUFACTURING RESOURCE PLANNING is a more complex form of back-scheduling, in which inventory, demand, a sales forecast and the priorities of management are all taken into account to generate a master production schedule. A computer can then generate detailed production schedules for thousands of parts, materials and processing needs. The success of the system depends on fast and accurate feedback on the status of parts, materials and other resources that must meet the back-scheduled deadlines in order to conform to the master schedule. The flow chart indicates the ideal flow of information in such a planning system.

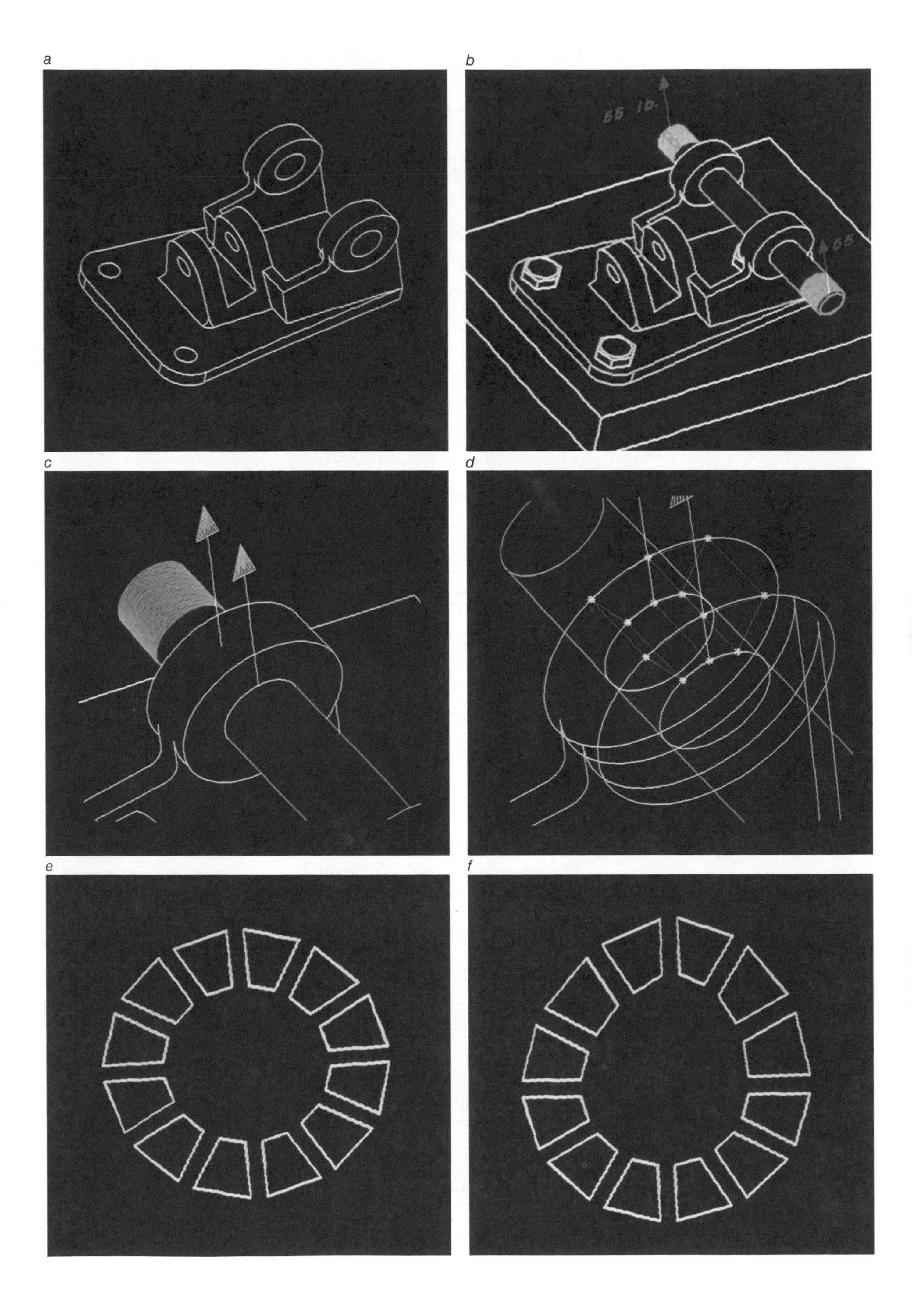
a
b
55 lb.
55
c
d
e
f

the design and the manufacturing departments. Because the part can be viewed in any orientation, at any scale and in any cross section the intent of the designer is much clearer than it is in a three-view drawing. In return one of the most important benefits to the designer is that an engineering analysis can be done quickly enough for several alternative design solutions to be tried. Engineers can analyze the part for its response to various kinds of stress without building a model or a prototype.

The engineering analysis is also done with the aid of a computer at a cathode-ray-tube terminal. In one method, called finite-element analysis, the part is divided into many small elements, or cells, and the response of each element to stress is observed on the screen. For example, the computer can generate an image of the part as it would look if it were deformed by mechanical stress, showing where the weaker regions are found. Other properties that vary with position, such as thermal and electrical conductivity, can be indicated with a color code for each cell.

The application of computer-aided design generally improves productivity in the drafting room by a factor of three or more, and it has brought striking overall benefits to manufacturers. At General Motors, for example, the redesign of a single automobile model required 14 months instead of the usual 24. Another company reduced the time needed to design custom valves from six months to one month. A manufacturer of molds for plastic parts was able to increase its output from 30 mold cavities per year to 140, solely because of the increased efficiency afforded by a computerized design system. Furthermore, the greatest savings that result from computer-aided-design systems are often effected during the assembly of the final product: the higher quality of the component parts makes the assembly faster and easier.

The information stored in a computer that specifies the geometric design of a part and the stages in its manufacture need not be limited to the part for which it was originally intended. In order to design a new part and plan how it will progress through the factory it is convenient to refer to a design and a process that are already established for a similar part. The need to identify such parts quickly can be met by a rationalized system for the storage and retrieval of information about parts, a system called group technology.

Group technology is in effect an electronic card file listing every part a company manufactures, together with a system for sorting the cards according to various characteristics of the parts. The parts can be classified in any way the company considers useful; generally parts are coded for such physical characteristics as size, shape, volume and materials used and for such manufacturing-process characteristics as the time required for the setup of the machinery, the machining sequence and the number of parts ordinarily made in a single lot. Once the parts have been classified the process planner for a new part can retrieve a list of old parts that have some of the same characteristics. He can then plan for the production of the new part simply by specifying that the manufacturing process is to be the same as that for the old part, with any differences noted. The procedure is called variant process planning.

The labor savings made possible by group technology is remarkable. Analysis has shown that in many companies only 20 percent of the parts initially thought to require new designs actually need them; of the remaining new parts 40 percent could be built from an existing design and the other 40 percent could be created by modifying an existing design.

Group technology can be applied not only to planning but also to the production machines themselves. Production machines can be grouped according to the parts for which they are employed; they can also be sorted into small cells of machines, each cell being dedicated to the production of a single family of parts. The regrouping allows a higher production rate and a more efficient use of the machinery.

Allocating the resources of a factory to maximize profits or productivity would be an exceedingly difficult mathematical problem. The methods of queuing theory and linear programming would have to be applied to a situation in which there may be hundreds of machines and workers, thousands of potential products and an almost unlimited number of routes a given product might follow during production. In the factory today, however, the practical problem is not to determine the best-possible configuration of labor, machines and products. Typically the organization of production is so far from the mathematical optimum that even a clearly suboptimal solution may offer substantial improvement. The immediate need is for a relatively simple method of planning and control that can cut down on long waiting times and eliminate most of the costs associated with inventory.

There are now several ways the computer can assist in planning and control. The simplest method is called manufacturing resource planning, which seeks to predict the demand for each element in the manufacturing process at a given time. For example, a manufacturing-resource-planning program could indicate how many milling machines (and how many operators for the machines) are needed in a factory making several products that call for milling. The method is an outgrowth of a system introduced by IBM in 1968 for determining when certain materials are needed in manufacturing. The basic idea of manufacturing resource planning is that the scheduling of labor, materials, machine time and other resource elements that go into the manufacture of the product can be estimated by extrapolating backward from the delivery date for the assembled product. If the scheduling is done accurately, there is no need to maintain a parts inventory because of uncertainties in the demand for parts; instead each part can be manufactured just before it is needed.

Suppose a company wants to make 50 pruning shears for shipment on September 1. To determine how many wood handles must be made and when they must be ready the manufacturing-resource-planning system consults a structured bill of materials for pruning shears. It finds that for every pair of shears two wood handles are needed. The system then determines that it takes, say, a week to assemble 50 pruning shears and two weeks to make 100 handles out of wood stock. The wood supplier requires a week's notice for delivery, and so the system automatically generates an order for the wood on August 4, four weeks before the shears are to be ready for shipment. The system might also generate additional orders

DESIGN AND ENGINEERING ANALYSIS can be done with great efficiency on a computer terminal provided with specialized programs. The operator can interact with the terminal by entering data and commands at a keyboard, by employing a "menu" of special-purpose commands or by pointing to the screen of a cathode-ray tube (or to a surface in front of the screen that corresponds point for point with regions of the screen) with a stylus; in addition a few frequently used commands can be issued by means of a hand-held control panel. In the photographs on the opposite page the analysis of the forces on a device called a brace plate is demonstrated. A model of the brace plate is generated (*a*). The plate is then schematically bolted down, an axle is passed through its two collars and a force to be analyzed is applied upward (*b*). To determine the effects of the force on a collar, the axle and one of the collars are enlarged (*c*) and the collar is sectioned by planes (*d*) into a matrix of small regions. The regions are best visualized on the screen if each region is reduced in size by a small amount so that the regions separate. The shrinkage is intended only to improve clarity; it does not affect the analysis. The unloaded configuration of the ring is rotated so that its axis is perpendicular to the screen (*e*). The maximum distortion of the collar under the load is then calculated, and the result is projected so that the distortion along the applied force is 100 times the distortion in other directions (*f*). Images were generated by the Computervision Corporation of Bedford, Mass.

for wood to be kept in inventory, but the inventory would be maintained only at the level needed to cover uncertainties in the supply of wood; no reserve would be needed for uncertainties in demand.

In order to introduce manufacturing resource planning successfully a company must have accurate information on the parts needed for each stage in the assembly of a product, on the time needed for manufacturing each part (including not only the time spent actually working on the part but also the time needed for setting up machines, for moving the part from one operation to the next and for delays while the part awaits processing at each station), on the lead time needed for purchasing parts from suppliers and on the company's own inventory. Many companies have failed in their first attempt to set up a manufacturing-resource-planning system because of insufficient data on these elements.

Nevertheless, more than 100 systems for manufacturing resource planning have been developed, and they have been put into use at more than 10,000 manufacturing sites. Their effectiveness has been demonstrated most clearly in factories where a considerable variety of products are made in comparatively small quantities; in these circumstances maintaining a large inventory would cut deeply into profit. Inventory reductions of up to a third and reductions of up to 6 percent in the cost of purchased parts have been achieved.

Manufacturing resource planning works quite well in job shops, where many parts are manufactured in varying quantities. When manufacturing is more repetitive, however, a system developed by the Toyota Motor Co. Ltd. called the Kanban system may be even more effective. In the Kanban system the order for a part to be made at one station of a production line is generated only by the requirements of the next station on the line. A chain of orders from work station to work station is thereby set in motion by a single order for finished products at the end of the line. Every component of the finished product, such as an automobile, is pulled through the line by the chain of work orders exactly when it is needed. Thus unlike the manufacturing-resource-planning system, which depends on detailed, centralized planning of all subassemblies, components and raw materials and on efficient feedback from every work station, the Kanban

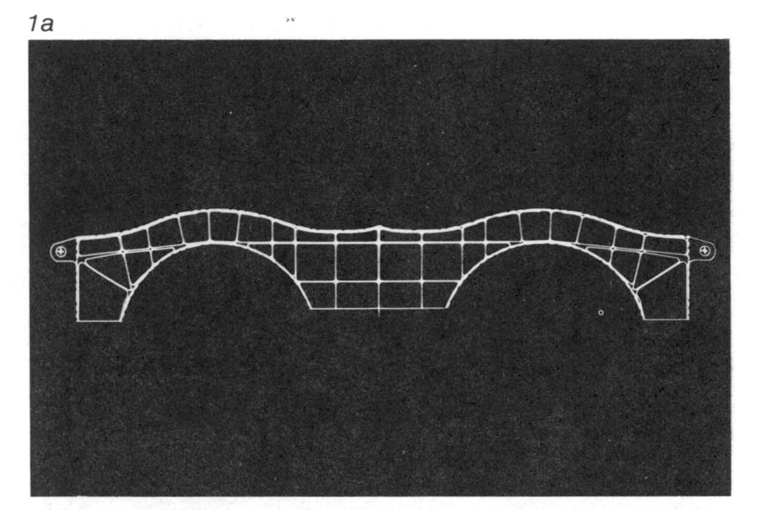

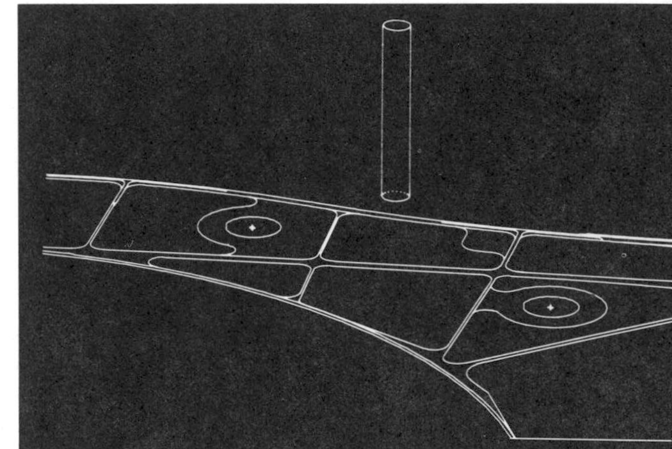

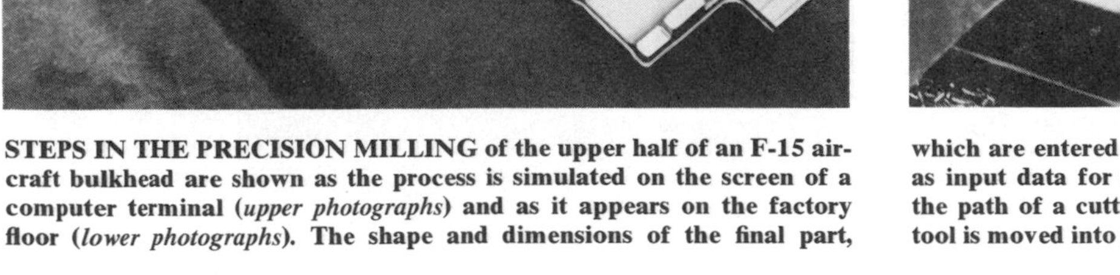

STEPS IN THE PRECISION MILLING of the upper half of an F-15 aircraft bulkhead are shown as the process is simulated on the screen of a computer terminal (*upper photographs*) and as it appears on the factory floor (*lower photographs*). The shape and dimensions of the final part, which are entered into the computer by the designer (*1a*), serve as input data for a computer programmer, who then specifies the path of a cutting tool on a rough forging (*1b*). The cutting tool is moved into place above the forging (*2a, 2b*) and enters the

system depends only on the centralized planning of the output of finished products. Moreover, in Kanban the parts are made in the exact quantities needed for production, with no allowances for wastage or spoilage, a feature of the system that requires standards of quality control now met primarily in Japanese factories. In the future computer programs will probably incorporate the best features of manufacturing resource planning for job-shop production and of the Kanban system (to the extent that non-Japanese societies can incorporate its methods) for repetitive production.

The basic principles of manufacturing resource planning can be extended in a number of ways. The generation of timely part orders can be based on the date the finished product must reach a certain warehouse or distribution center instead of on the shipment date. This method of scheduling is called distribution resource planning, and it must take into account the time needed for shipment. It can be employed to generate shipment dates for various products; its output can become input for a manufacturing-resource-planning system.

The computer can readily bring together information from a manufacturing-planning or a distribution-planning system to generate summary reports for the management of a company. The reports could include the overall backlog of orders, the inventory level, the daily production rate and the daily difference between the input to the plant and its output. If the management wants to consider alternative production rates, inventory levels and the like, the computer can rapidly simulate the consequences of the changes for the rest of the company. Once a production plan is chosen it can still be optimized by the more rigorous mathematical techniques of linear programming and queuing theory.

A common requirement for all versions of resource planning is feedback about operations on the shop floor. Information on the movement of material, the performance of workers and machines and the attendance of workers can be collected by various means. For example, a worker's time card can be imprinted with a machine-readable code such as the Universal Product Code bars, so that the working hours recorded by the time clock are automatically assigned to the worker. The papers that accompany a job through various stages of processing can be similarly

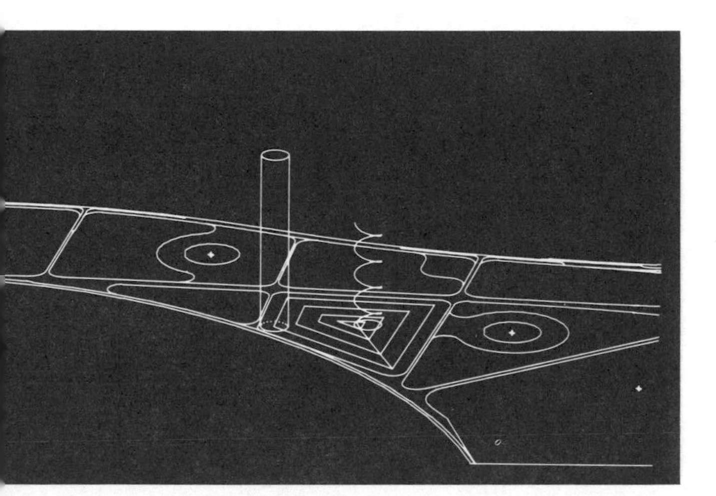

4a

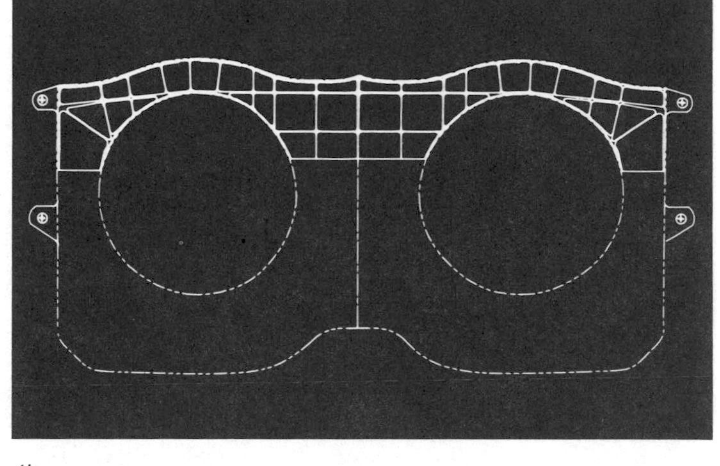

4b

workpiece along a spiral path; it then cuts an ever widening region in the forging (*3a, 3b*). The operation of the milling machine is controlled by the computer program. There are three cutting heads mounted in parallel, so that three bulkheads can be machined at one time. Each cutting head of the machine can move along three axes in space and can tilt around the two horizontal axes, cutting to a tolerance of a ten-thousandth of an inch. The workpiece is attached to a similar machined part to create the finished bulkhead (*4a, 4b*).

encoded, or the product itself can be. Typed messages can be entered from computer terminals throughout the factory. The information enables managers to determine whether a part is meeting the schedule set for it by the planning system, and if it is not, to decide what measures should be taken. The information feedback need not be registered in the computer immediately; in most circumstances an update once or twice a day is sufficient.

One of the most important benefits of a system for manufacturing resource planning and control is that it enables the company to respond quickly to changing market conditions. Before the introduction of automatic planning systems the response to changing priorities was the duty of the expediter, and the role of expediter usually fell to the shop foreman. Consequently when such systems are installed, the foreman can go back to being a foreman, that is, he can focus on being the leader of a group of workers rather than spending his time in efforts to relieve shortages of parts, to order repairs for machines and to shepherd the latest top-priority item through the production line.

Although I have emphasized changes in the organization of the manufacturing enterprise, the mechanization of operations on the factory floor is also continuing. Data-processing technology can be applied to the control of three general kinds of machines in the factory: machines that store, retrieve or transport materials, machines that process the materials and robots.

Automatic storage and retrieval systems transfer pallets of material into or out of storage racks up to 100 feet high. Smaller systems called miniloaders hold drawers of small parts. In both cases a part can be selected by number or by location, and an automatic shuttle is actuated that retrieves the part. Essentially such a system is an automatic warehouse in which the shuttle takes the place of the fork-lift truck and its human operator. Similarly, automatic guided-vehicle systems replace conveyors and hand trucks for transporting materials to and from the warehouse and throughout the factory. The driverless shuttle cars can be guided by signals sent through a wire embedded in the floor.

The earliest numerically controlled machine tools were programmed by means of a punched paper tape. Each instruction to the machine was repre-

5a

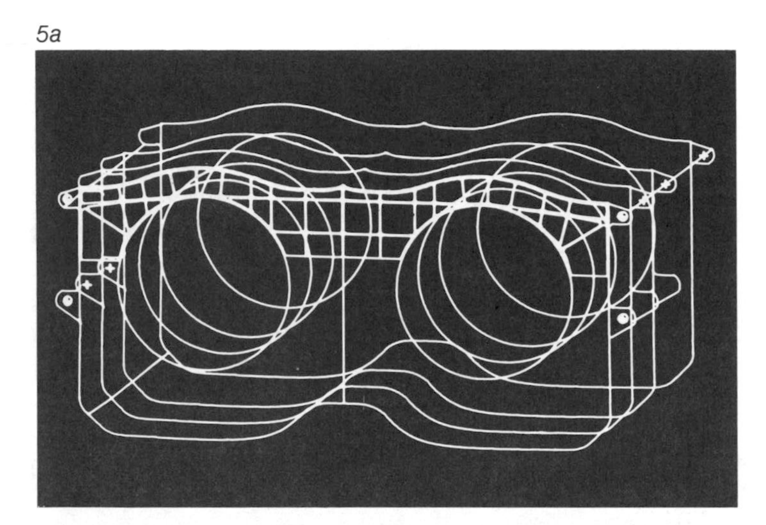

6a

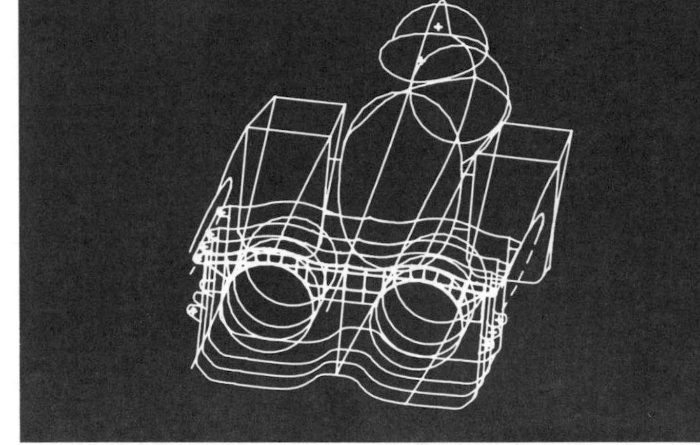

5b

6b

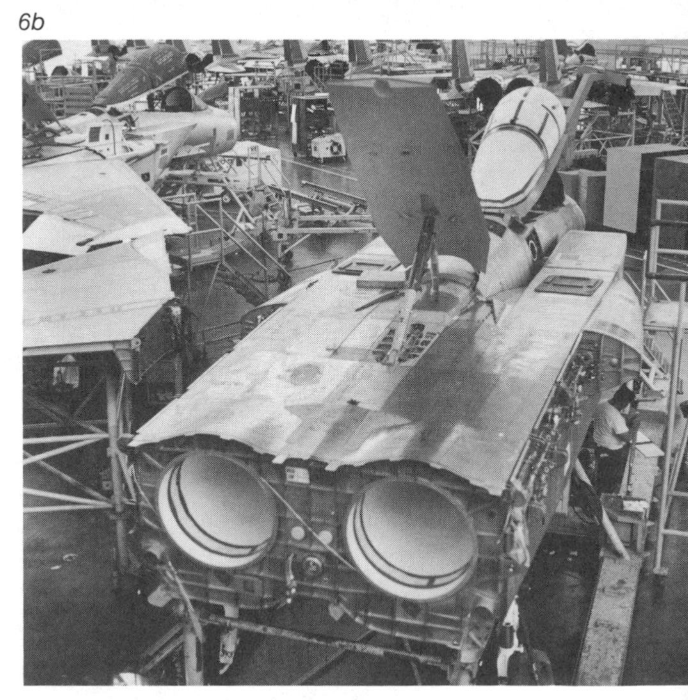

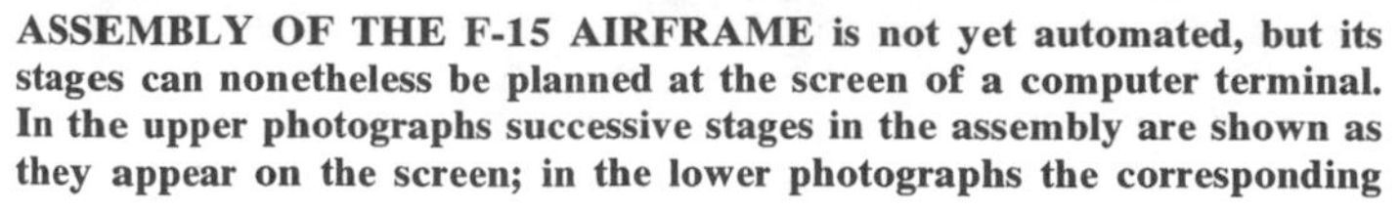

ASSEMBLY OF THE F-15 AIRFRAME is not yet automated, but its stages can nonetheless be planned at the screen of a computer terminal. In the upper photographs successive stages in the assembly are shown as they appear on the screen; in the lower photographs the corresponding stages are shown in the assembly plant. The bulkhead is first mated to three similar bulkheads to form the midsection of the aircraft (*5a, 5b*), the midsection is attached to the front section (*6a, 6b*) and finally the wings and tail section of the aircraft

sented by a pattern of holes in the tape; the pattern was decoded by an optical or mechanical reader attached to the tool. In most cases the paper-tape reader has been replaced by a small digital computer mounted on the machine. A modern computer-numerically controlled tool can be as big as a small house and can incorporate a cutting head capable of independent motion around several axes at the same time; the computer control enables the machine to cut metal automatically to tolerances of a ten-thousandth of an inch. Moreover, the program can prevent the machine from cutting too deep into the workpiece and so ruining the part, and in some cases it can signal the machine operator to change or sharpen the cutting tool when sensors indicate that the torque required to make the cut is outside the proper range of values.

When several computer-numerically controlled machine tools are linked by a hierarchy of computers, they are called direct-numerically controlled machine tools. Typically each machine is controlled by a microcomputer; several machines are linked by a minicomputer, and several minicomputers are tied in turn to a large mainframe computer. The programs for the manufacture of every part the company makes can be stored in a central data base, and they can be transferred from the mainframe computer to any of the machine tools in the network. In addition information about the status of each machine, the volume of its production and the quality of the finished parts can flow back to the mainframe computer from the peripheral controllers. As many as 100 machine tools can be connected in such a hierarchy.

In a direct-numerically controlled system the only connection between the tools is electronic; the workpiece must still be moved from one machine to another by manual methods. If several direct-numerically controlled machine tools are further linked by a materials-handling system and the mainframe computer is programmed to operate the tools in a specified sequence, the result is called a flexible manufacturing system. In such a system families of parts are selected through group technology for machining. Once a pallet of workpieces is set in place the workpieces proceed automatically from tool to tool, where they are machined in the proper sequence. The entire system may require loading and unloading only once a day; one person is needed to oversee the operation of the system for the rest of the day. Furthermore, the fraction of each shift a machine spends cutting metal can be as high as 50 to 90 percent in a flexible manufacturing system; with a computer-numerically controlled machine tool standing alone the cutting time may be as low as 10 to 30 percent of the total shift time.

The higher the level of integration among machines, the greater the need for some form of automatic inspection of products. A worker operating a machine tool manually can note a defect and stop work immediately, but a machine running autonomously could, through a mechanical failure or a programming error, ruin an entire batch of parts. Information from various sensory devices on the machines can be employed to accept or reject individual parts. The information can also serve to build a statistical data base. The statistical summary is required in certain industries, such as pharmaceuticals and aircraft manufacturing, and statistical feedback makes it possible for the computer that controls each machine tool to adjust the tool during production.

A manufacturing system that has become emblematic of factory work in general is the assembly line. It should be noted that the assembly line is not necessarily a mechanized system; it is a method of organizing work that can be applied either to human workers or to machines. Many products are still assembled by hand, with each worker doing one small step of the job and passing the workpiece on to the next station.

For products that are made in large quantities the assembly process can be automated entirely by building a single-purpose machine. The design and construction of such machines constitute a highly developed art that draws on a variety of ingenious methods for orienting parts, fitting them together and fastening them. In most instances

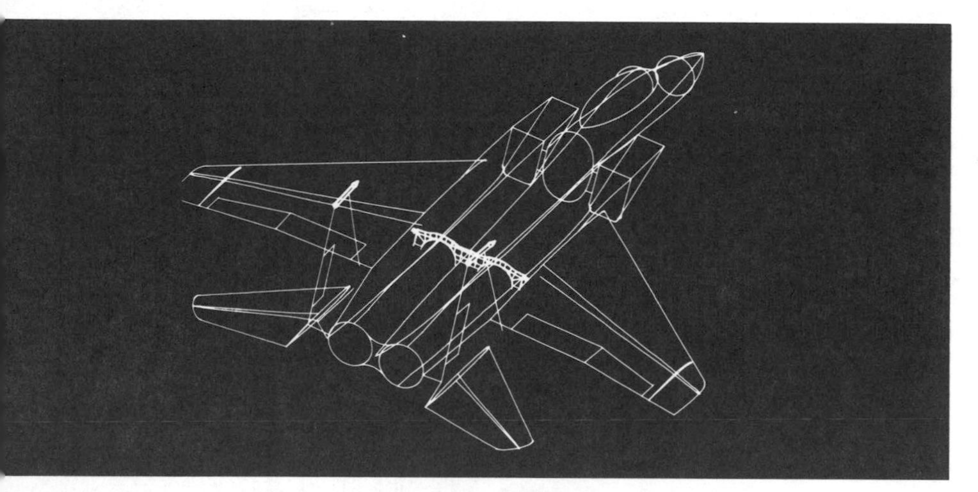

are added (*7a, 7b*). The turbine engines are installed during a later stage in the assembly. One advantage of the computer simulation of the assembly is that the designer can determine at a glance whether or not the components will fit together properly. The photographs on these two pages and on the preceding two pages were made at McDonnell Douglas in St. Louis.

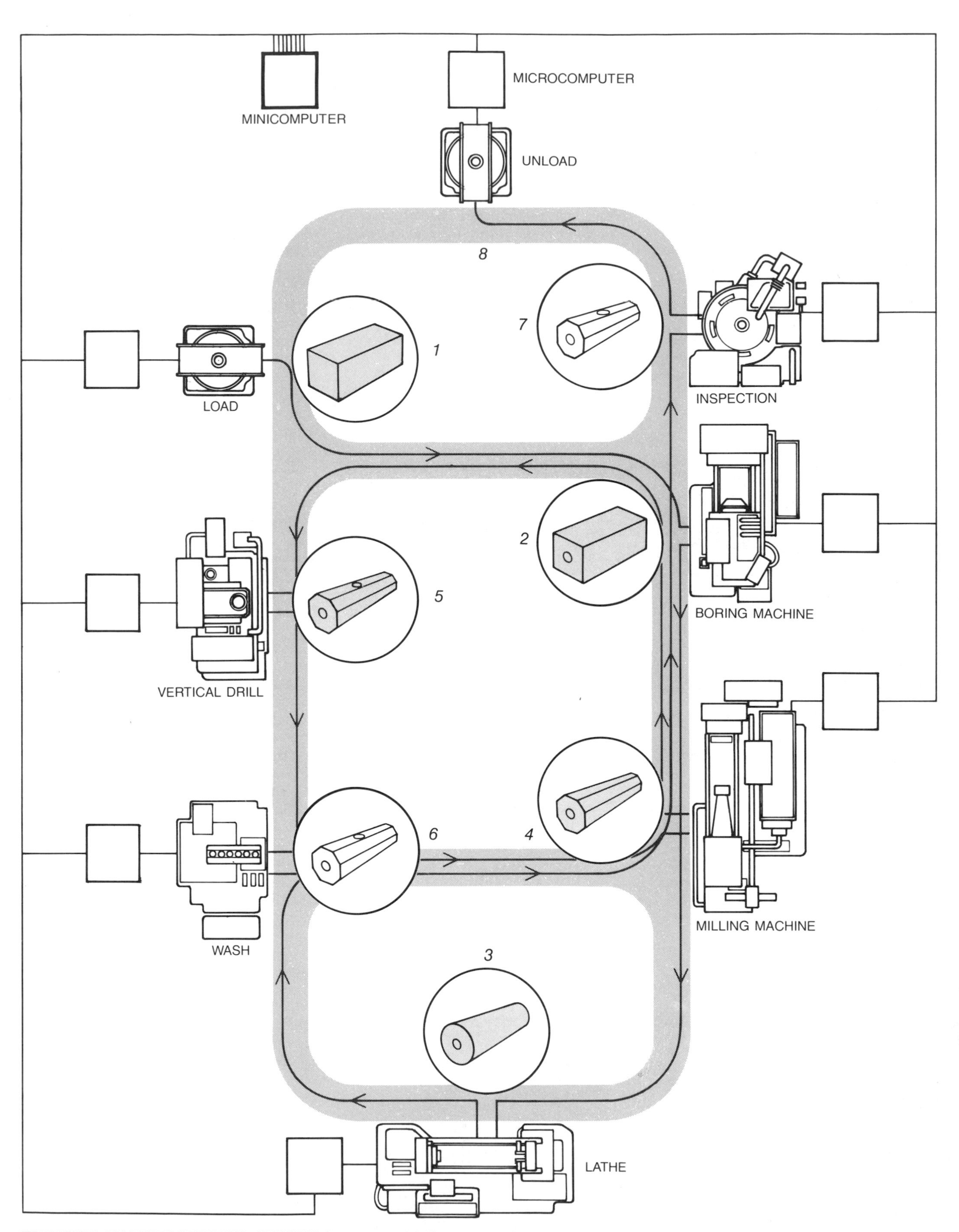

FLEXIBLE MANUFACTURING SYSTEM is an automated set of programmable machine tools for metalworking. The machines are controlled by a hierarchy of computers and are linked by a conveyor that carries workpieces from one machine to the next. The minicomputer determines the overall sequence of operations to be carried out on each workpiece. When the workpiece reaches a machine, the minicomputer also directs the machine to select a cutting tool and "downloads" a program into a smaller microcomputer that controls the cutting path of the tool. Flexible manufacturing systems have now been built that can run for hours without intervention. Parts to be machined are loaded at the entry to the system during the first shift, and the system operates throughout the second and third shifts. Setup times are so reduced that such a system may be able to manufacture 100 randomly selected rotational parts in 72 hours.

the design of the product itself is modified to facilitate automatic assembly. The chief disadvantage of such "fixed tooling" is a lack of versatility: a machine for the assembly of fountain pens cannot be adapted to the manufacture of ballpoint pens when the demands of the marketplace change.

The robot, a programmable machine capable of moving materials and performing repetitive tasks, is beginning to make automated assembly economically feasible in some lower-volume applications. In certain cases the robot replaces a human worker carrying out some routine operation, such as loading products onto a pallet. In other cases a system of robots is a more flexible (but generally slower) alternative to fixed tooling.

The main difficulty in the use of robots for assembly is that the robot is not yet able to pick a randomly oriented part out of a bin. If the orientation of the part is preserved at all stages of the assembly process, however, the robot can compete economically with other machines. One of the most important applications of robots in the U.S. is in the loading and unloading of machine tools. The other primary applications so far are in jobs that are dirty, hazardous, unpleasant or monotonous. Between 5,000 and 7,000 robots are currently used in American industry for spot welding, spray painting, machine loading and unloading and certain assembly operations. In Japan the robot population is about 80,000, but the Japan Industrial Robot Association accepts a broader definition of robots, including simple mechanical manipulators with mechanical stops that would not be considered robots in the U.S.

Although robots and computer–numerically controlled machine tools are alike in being programmable, the robot is generally much smaller and can easily be moved about. Moreover, in many cases the robot is programmed analogically, that is, by placing the device in the "teach" mode and moving its arm exactly as the job demands. Hence many robots act as recording and playback devices that simulate human motions, although they can also be programmed with a set of coded instructions in a high-level computing language. Their main advantage over human workers is that their performance never varies; they are rarely faster than human workers, but they never tire and they are often more reliable.

The integration of the six major areas of manufacturing technology—design, group technology, manufacturing resource planning and control, materials handling, manufacturing process machines and robots—depends on a carefully designed hierarchy of information flow. The linkage of the six areas is brought about by a centralized processing system, but there are also many kinds of information in each of the areas that need not be centrally stored. Distributed, or decentralized, data processing has become possible in the past 10 or 15 years because of the tremendous growth in microelectronic components and the decreasing cost of storing and manipulating information. In manufacturing, as in other sectors of the economy, the trend is to disperse powerful minicomputers and microcomputers to the individual workers, making each worker responsible for data entry and processing control. Computers can then be linked to one another and to the central data base of the company by telephone lines or another telecommunications network.

Given the capabilities of existing technology, it is possible to imagine in some detail how a factory could operate if all six of the areas I have discussed were linked in interdependent modules. The factory floor would be divided into cells defined by their manufacturing function, such as a design cell, a flexible-machining cell, a welding cell and an assembly cell. Dozens of robots might be linked by a hierarchy of computers, much as direct–numerically controlled machine tools are today. Feedback to

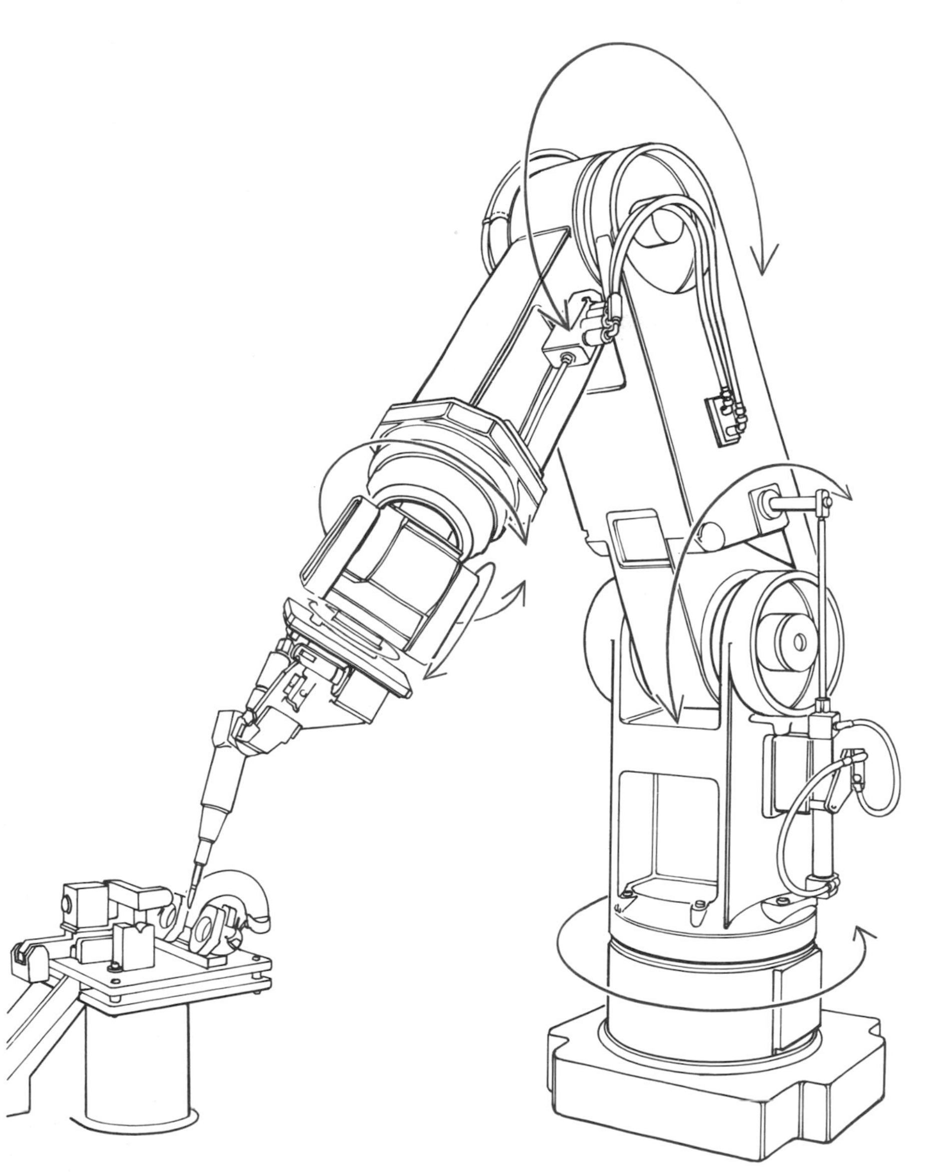

ROBOT ARM capable of independent movement around six axes is a freestanding machine that is programmed analogically. An operator "teaches" the robot a three-dimensional pattern of motions by moving the end of the arm through a sequence of positions and orientations. The robot records the maneuvers in the memory of a computer and repeats them indefinitely when it is called on to do so. The end of the arm can be fitted with a grasping device or another tool, such as a paint sprayer or a pair of spot-welding electrodes. The robot is shown with a spinning abrasive tool for deburring castings. The robot has no sensors, but several other kinds of robot have been equipped with devices that sense torque, gripping force, the visual field and other features of the environment. It has not yet been possible, however, to endow a robot with the ability to pick a randomly oriented part out of a bin of similar parts. Hence, unless the orientation of a part is constrained before the robot encounters the part, the adaptability of the robot to assembly and other tasks that require the exact orientation of the part is limited. The illustration shows the Type 80 vertical robot manufactured by the Ateliers de Constructions Mécaniques et Automation (ACMA) of Beauchamp, France, a subsidiary of Renault.

the manufacturing control systems from the robots, from the machines and from the people in the factory would be immediate, and so the planned flow of products through the factory could be adjusted continuously to reflect changes in operating conditions. The design of the plant would emphasize flexibility, so that a variety of products could be made by the same machines; indeed, products might be made in unit quantity.

Communications between the factory and a company's most important outside customers, suppliers and subcontractors would be carried on directly among the computers of the various organizations. Drawings, for example, would not be issued to a subcontractor; instead the geometric data and machine-tool programs needed to

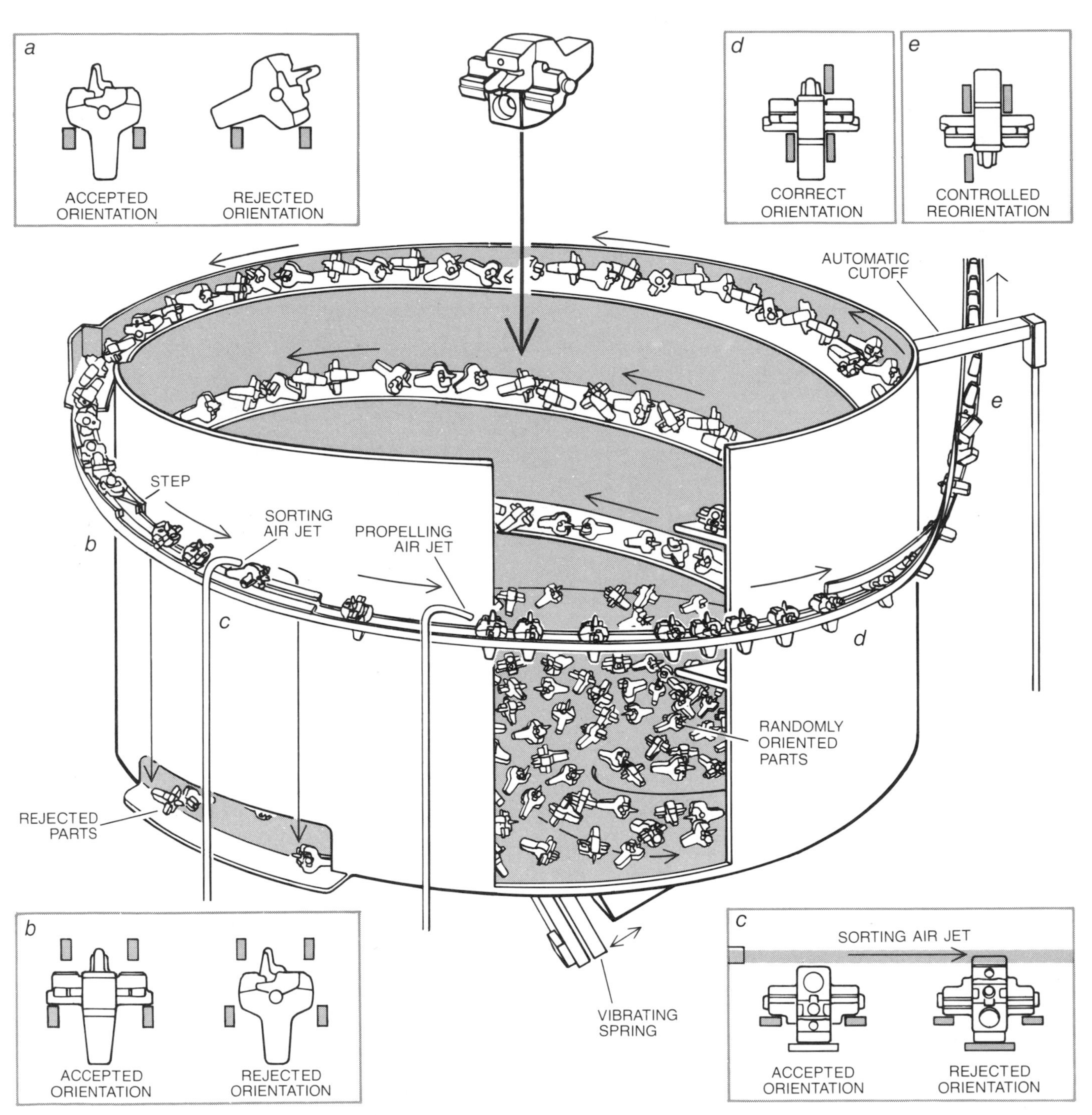

CORRECT ORIENTATION OF PARTS is a prerequisite of automatic assembly, but it is one of the most difficult processes to mechanize. For small, lightweight parts a device called a vibratory feeder can do what a robot cannot: it can reject all configurations of the part except the one needed for automatic assembly. In the illustration a bin of randomly oriented toggle switches is mounted on a system of springs and vibrated up and down at a rate of between 60 and 120 times per second. The vibrations impart a twist or torque to the bin, with the result that the switches "walk" up a spiral ramp on the inside of the bin. At the top of the bin the ramp meets a two-rail track, and the switches either fall off the track and back into the bin or continue to move along the track with the toggle pointing down between the rails. Only one of the four possible orientations can be accepted, and so the switches are sorted at two stations along the track. At one station they move down a series of steps in the track and the vibrations cause switches in two of the four orientations to fall back into the bin. At the second station the rest of the wrongly oriented switches are blown back into the bin by a jet of air. The device can feed correctly oriented toggle switches to the assembly machines at a rate of 2,400 per hour. Vibratory feeders can be constructed for orienting almost any small part, but for parts longer than about six feet or heavier than about half a pound the rapid wear of the vibrating parts makes such a finely tuned system impractical. The machine in the diagram is manufactured by the Bodine Corporation of Bridgeport, Conn.

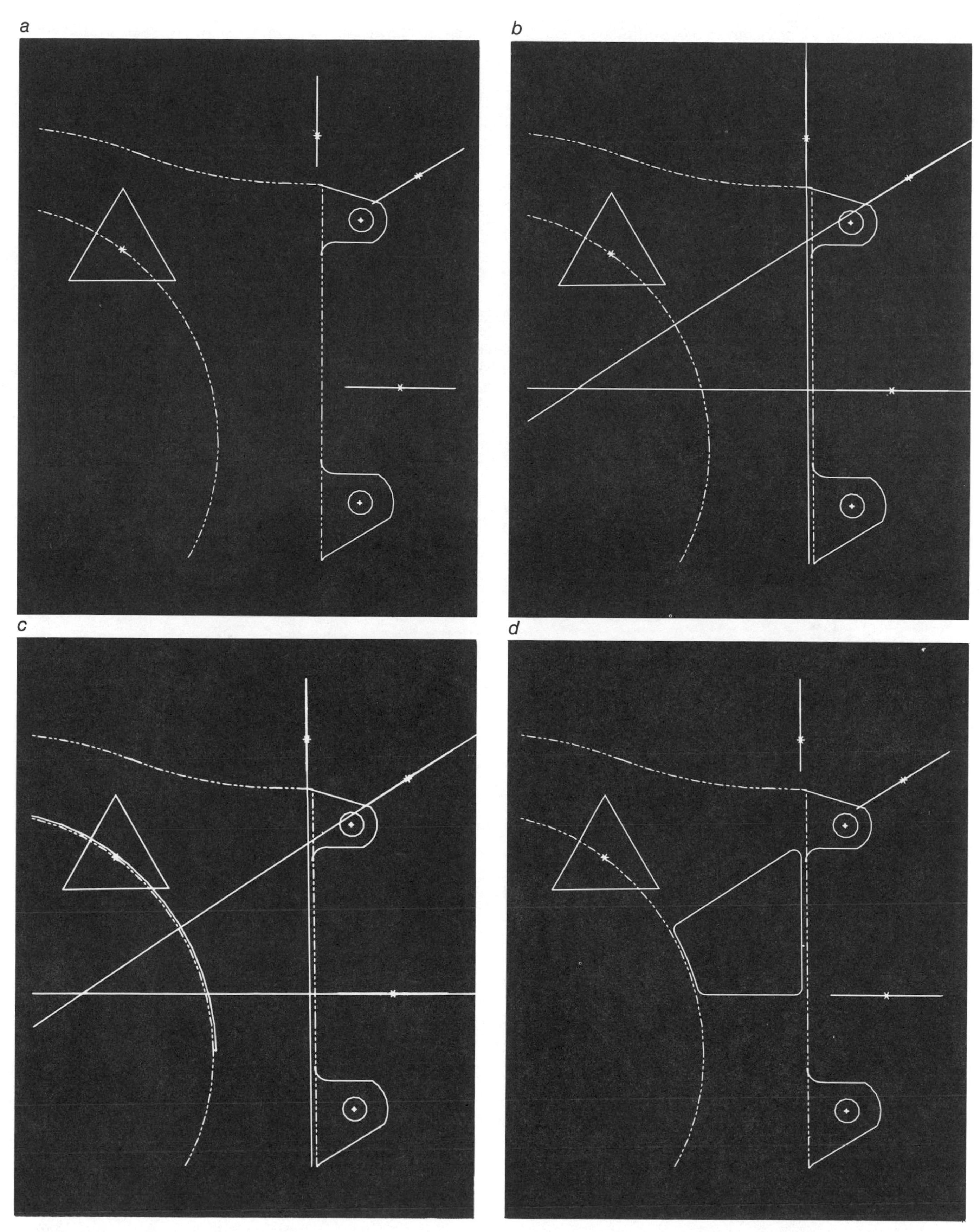

COMPUTER-AIDED-DESIGN TERMINAL enables the designer to create complex shapes by combining simpler ones. In the photograph at the upper left (*a*) a computer terminal displays the basic outline of part of the upper section of a bulkhead for the F-15 fighter aircraft. Three planes perpendicular to the plane of the screen have been defined by three line segments, each of which is marked by an asterisk. The orientation of the plane of the bulkhead is indicated by a triangle with an asterisk; when the triangle is equilateral, as it is here, the bulkhead is parallel to the plane of the screen. In the second photograph (*b*) the lines formed at the intersections of the three planes with the plane of the bulkhead are shown. In the third photograph (*c*) the computer has drawn a curve slightly offset from a curve already drawn. In the photograph at the lower right (*d*) the closed region defined by the curve and by the three lines is automatically given rounded corners. The dimensions of the region, although they are not shown on the screen, are stored by the computer for retrieval when needed for engineering analysis or process design. The images were generated by the McDonnell Douglas Corporation in St. Louis.

shape a part would be transferred electronically to the subcontractor's computer. Similarly orders from major customers or to major suppliers would be transmitted electronically. Within the company separate divisions would be interconnected by a satellite-based communications system.

The size of the individual manufacturing plants is likely to be small: fewer than 500 people per plant in most cases. For the manufacture of some types of products the size of a plant could be as small as five workers. The Yamazaki Machinery Corporation of Florence, Ky., is setting up a plant for the manufacture of machine-tool parts that will employ only five workers on the first shift, one worker on the second shift and one on the third. Such small plants would create a more personal working environment, and the efforts at each plant would be focused on the manufacture of a few families of products. No single plant, however, would have much influence on the policies or financial well-being of the company as a whole.

The reduction in the number of blue-collar workers in factories is likely to continue as robots and flexible manufacturing systems are installed. Many other trends in manufacturing, however, will affect skilled professional workers as well as craftsmen and laborers. Mid-level white-collar workers will probably have to be retrained to exploit the new technology, and engineers will find it necessary to continue their education. Hence it is likely that corporate design centers will continue to be attracted to the environs of cities with major educational centers, such as Boston and San Francisco. Independent engineering service companies will probably establish themselves near the same cities.

Given the undeniable dislocations in the work force and the expense of information-processing technology, why would one expect the technology ever to be installed? For the short term the answer most often given is that manufacturers must compete in worldwide markets. Manufacturers who acknowledge that information technology will put people out of work argue that without the technology they would not be able to compete at all, and all of their employees would lose their jobs.

The mechanization of design and manufacturing promises the manufacturer higher productivity, better quality at lower cost, the ability to give better customer service and the flexibility to meet the demand for an increasing array of products and options that have shorter life cycles than ever before. In sum, information-processing technology will continue to revolutionize the way work is done in design and manufacturing. Information and the ability to transmit it quickly will come to be recognized as a resource as valuable as money in the bank or parts on the shelf.

5

THE MECHANIZATION OF COMMERCE

The Mechanization of Commerce

by MARTIN L. ERNST

Such services as finance, transport, distribution and communications are being mechanized even more than the production of goods. In the process they call for workers with ever higher levels of education

Commerce encompasses all the economic interactions among the members of a society. It is information-intensive: it requires that information be made available on goods, their prices and their utility, and the modern means of paying for goods requires that financial information be transmitted between the parties involved. Moreover, the storage and transportation of goods call for detailed records such as schedules and manifests. Commerce has therefore been quick to adopt new forms of technology for processing information. Clay tablets found in the Middle East furnish strong evidence that writing originated with commercial records. More recently the sales history of computers shows that the large-scale industrial application of computers began with commercial functions.

Today, when commerce is responsible for more than 35 percent of all employment in the U.S., the new electronic technologies dominate the mechanization of financial institutions. In other commercial sectors nonelectronic machines play a larger role. Beyond this some central themes emerge that are useful for understanding the mechanization of commerce: how it arises, how it is implemented and what its impacts are likely to be. Here I shall give a series of examples to develop these themes. The examples will come from four major areas: finance, transportation, the distribution of goods and lastly communications.

One theme concerns the role of government in commerce. Traditionally governments have performed certain commercial functions or have intervened in them. Good coinage, honest weights and balances, well-maintained roads and port facilities and equitable (or at least consistent) judicial systems played a major part in establishing the success of city-states and later of nations. Hence most aspects of finance, transportation and communications in the U.S. have been subject to considerable regulation, and the first-class mail service is operated by the Government as a monopoly. In many countries the same activities are conducted almost entirely by government organizations, and in virtually all countries a major part of the legal structure is devoted to codes concerned with commercial transactions. Inevitably, then, government intervention has influenced the adoption of new technology in commerce, sometimes favorably, sometimes not. The basis on which an industry is subsidized or regulated establishes incentives that favor the rapid adoption of some technical applications and the slow adoption of others.

A second theme is the almost universal requirement in commerce for interactions among independent parties. Banks must cooperate with one another, and moreover they must have a considerable degree of uniformity in their basic procedures if checks are to be a useful way to pay bills. Transportation companies, wholesalers and warehouse operators must work together to ensure the flow of goods from factories to stores. Even the U.S. Postal Service and the Bell System, which hold a dominant position in their field of communications, must rely on others for significant parts of their total function. For example, the Postal Service relies on airlines and on the manufacturers of sorting machines. The need for cooperation often limits individual organizations in their choice among available technologies. Often a move toward mechanization can be made only after broad agreement within an industry or across several industries, and this requirement can dictate the pace and the nature of change.

In the light of this second theme the third is ironic. Many of the most effective mechanizations in commerce have been based on quite modest devices. Many others have been based on new configurations of standard machines. Often they are machines that have long been used in other economic sectors.

Among financial industries banking offers the widest scope for examining the nature and impact of mechanization. A good place to begin is with a familiar item: the checks people and businesses employ to receive and make payments. Offered by some 14,000 commercial banks (and now by some 5,000 savings banks through Negotiable Orders of Withdrawal, which are the equivalent of checks), checks and their processing have always been labor-intensive and form an obvious target for mechanization.

Regardless of where it is deposited, a check, or at least the information on it, must be returned to the person who wrote the check by way of his bank. The journey is sometimes a long one. The mechanization of the process began years ago with the introduction of electromechanical check-sorting equipment and bookkeeping machines. Then the Federal Government began producing almost all its checks on punch-card stock. These early steps, however, were fairly limited. The start of modern check processing was the introduction of magnetic-ink character recognition (MICR) encoding on all checks. These machine-readable numbers identify the account on which the check is drawn and the bank in which the account is held. Complete machine processing becomes possible when numbers that identify the receiving account and the bank are entered into computerized records by the bank at which the check is first

CONTAINERIZED SHIPPING exemplifies the application of an ancient and fundamentally simple technology, containers, to the mechanization of modern commerce. Here containers are arrayed on the dock of the Global Marine Terminal in Port Elizabeth, N.J. Most of them are 20 feet long and hold from 8.5 to 11 metric tons of goods. The ship is the *Neptune Diamond,* out of Singapore, which makes a circuit of five U.S. ports and five ports in the Far East. It carries a maximum of 2,100 20-foot containers, about a fourth of which are off-loaded at the Global Terminal and replaced by other containers during a stopover of some 30 hours.

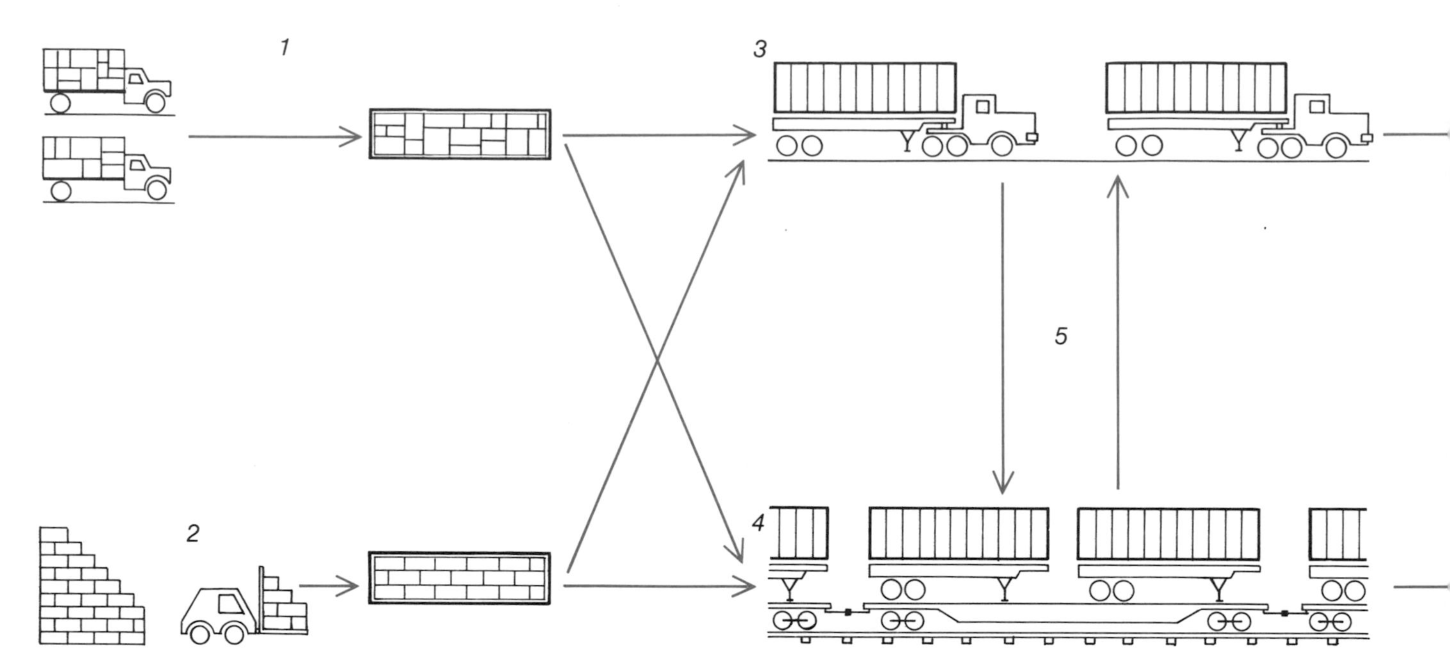

MOVEMENT OF CONTAINERS serving a transoceanic shipping route is diagrammed. Goods from small shippers are packed in containers at local terminals by consolidators and freight forwarders (*1*); goods shipped in large quantity (*2*) need no such mediation. The containers are carried overland by truck (*3*) or train (*4*); the trains may be organized specifically to move sets of containers. In some instances containers are transferred from one mode of transportation to another (*5*), a circumstance favoring the growth of transportation compa-

deposited and the amount of the check is MICR-encoded on the check itself.

The introduction of MICR took almost 15 years. It began in the early 1950's with a series of studies sponsored by several bank associations and the Federal Reserve System. In 1958 standards for coding, the position of codes on checks, an acceptable printing ink and similar requirements were established. Another 10 years were to pass before essentially all banks employed MICR, but with the larger banks taking the lead some 85 percent of all checks were being encoded with MICR numbers by 1963. The process illustrates the long time it takes when a large number of organizations must be brought together to make standard changes, even if the changes appear to be simple.

MICR's stream of machine-readable data speeded the development of computer-based record keeping for virtually all checking accounts. And the benefits of having mechanized account records spread rapidly to other bank activities. Computer-based equipment for tellers was introduced early in the 1960's to speed the entry of data and the processing of transactions that originate at a teller window. More recently the existence of computer account records has facilitated the growth of automatic teller machines (ATM's). These have become increasingly popular; cash withdrawals and certain other operations, such as transfers between accounts at a single bank, can now be done at any time rather than only during banking hours.

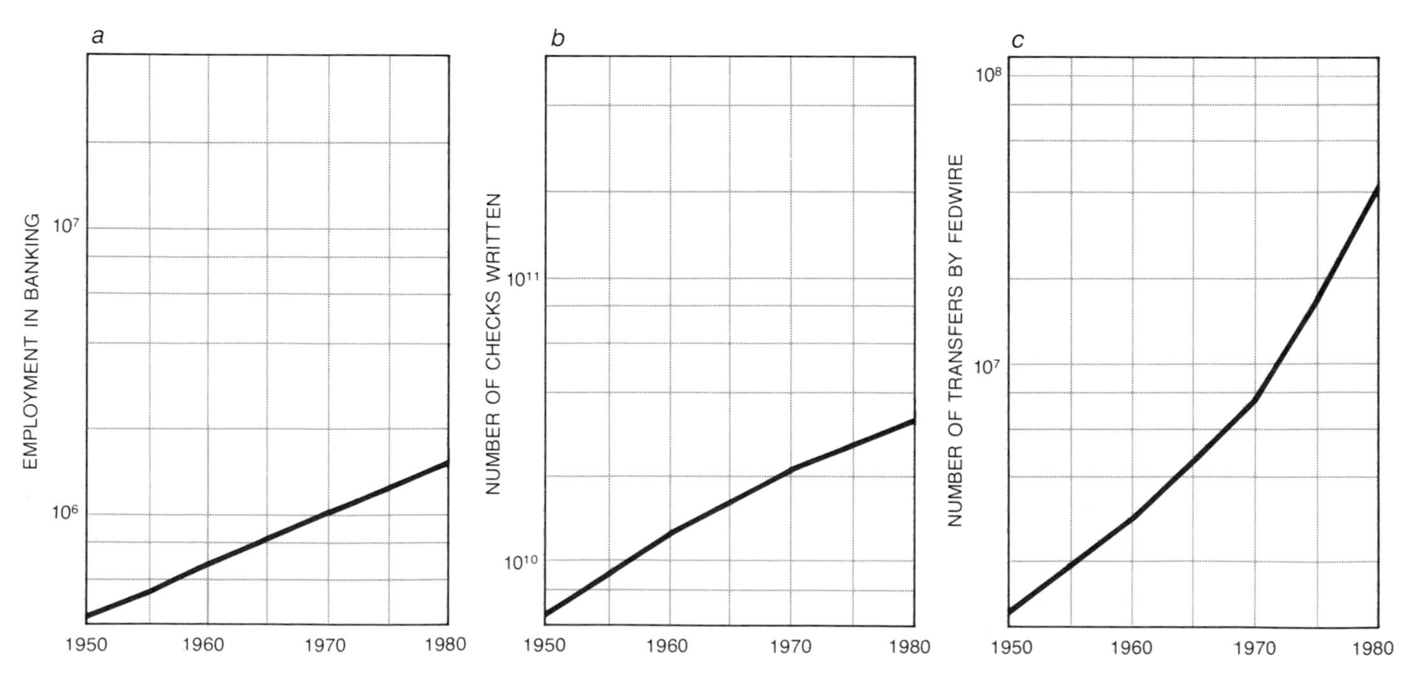

SIX INDICATORS OF GROWTH in the U.S. banking industry are charted logarithmically, so that equal rates of increase among the indicators would be represented by lines of identical upward slope. Employment in the industry (*a*) more than tripled in 30 years, and its share of the civilian work force more than doubled (to 1.5 percent in 1980). Meanwhile the number of checks being written (*b*) increased almost fivefold, the number of interbank transfers of funds handled by the Federal Reserve System's telecommunication network, Fed-

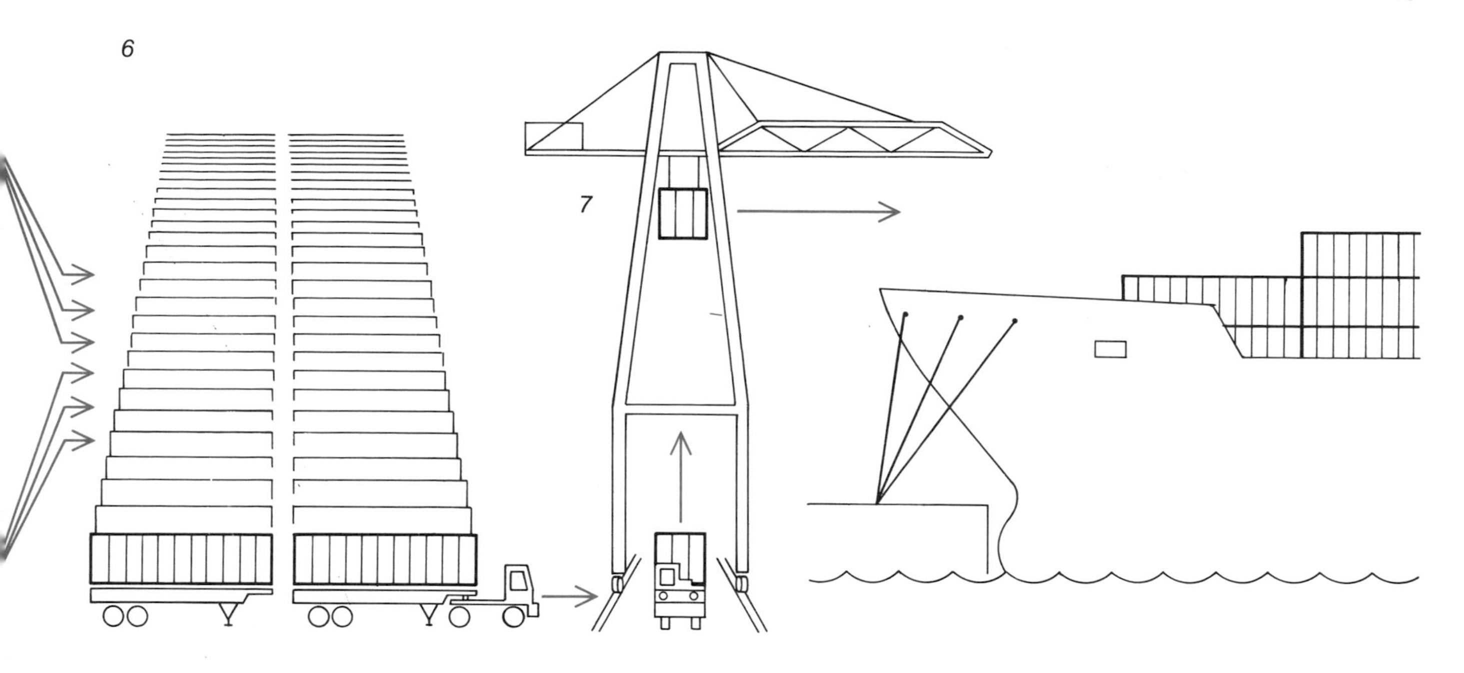

nies that operate in more than one mode. On the dock (*6*) the containers get the attention of longshoremen and customs brokers and agents; then they are taken by a specialized tractor to the crane (*7*) that transfers them to a ship. The wheel flatbeds that carried the containers are left behind. A counterflow, managed by computer, sends empty containers to where they are needed. Computers also keep track of containers' locations and transmit ships' manifests (cargo lists) from port to port ahead of the ships by telecommunications.

The ATM's illustrate a characteristic of much of the mechanization of commerce. In an ATM transaction the customer himself punches a set of codes into a terminal; thus he provides his bank with machine-readable data. In effect he is doing work for which the bank previously had to use its own staff. This type of "labor sharing" is becoming increasingly common. It is part of a trend that has been under way for decades; in the 1930's, for example, self-service in supermarkets began to replace what used to be the function of store clerks.

A logical next step in the evolution of check processing will be for payments to be made by means of an electronic terminal in the home. The terminals themselves are technically quite straightforward. Their installation nonetheless awaits several developments. First, the population of home terminals must be large enough to make a bill-paying service economically feasible. Second, enough banks and merchants must agree on the terms of the service to give it a reasonable degree of universality. Finally, consumers must have reasons to want to use the terminals.

Potentially there are incentives for all the participants in such systems. Viewed in their totality the costs of checks are not small. When a check is used to pay a bill, for example, the cost is about $1. Somewhat more than half of that cost is incurred by the banks, and the remainder is incurred by the biller and the payer for postage, paper and printing. The current costs, however, are almost invis-

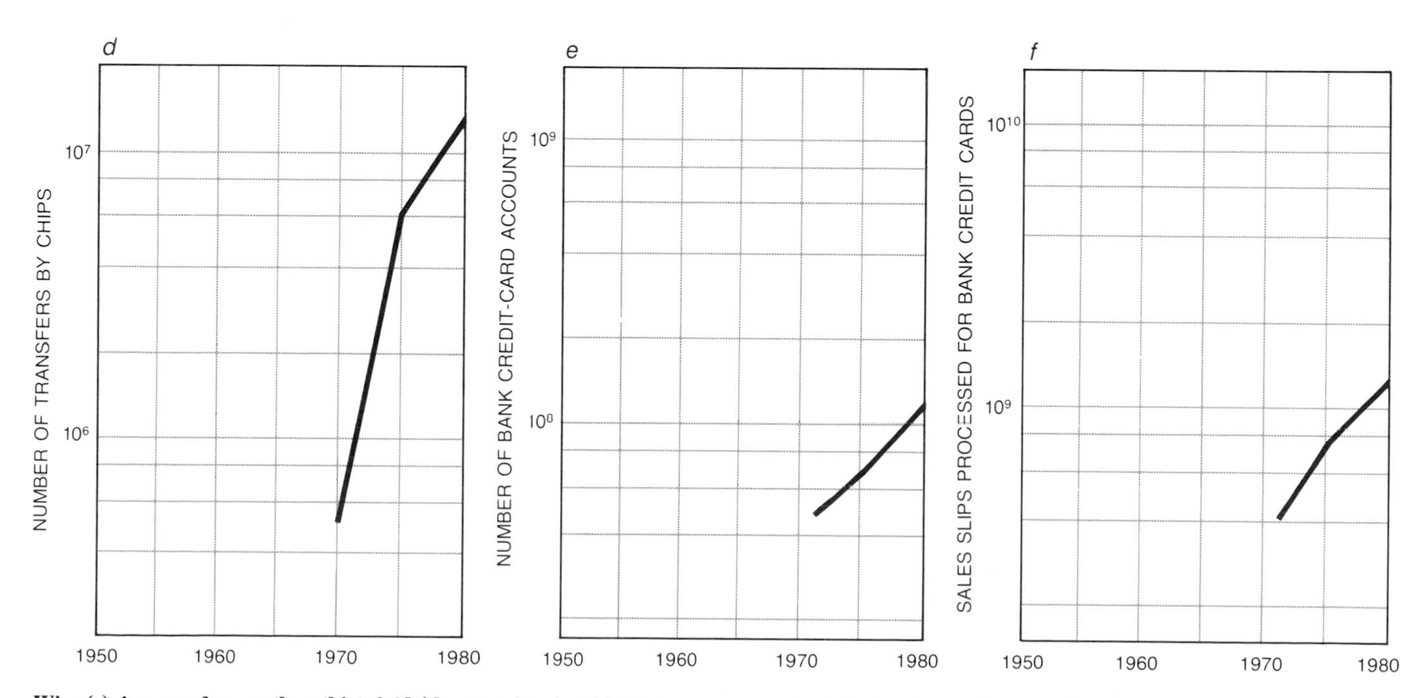

Wire (*c*), increased more than thirtyfold (the transfers in 1980 had a value of $78.6 trillion) and private transfer networks such as CHIPS, or Clearing House Interbank Payment System (*d*), began operation. In addition bank-issued credit cards (*e, f*) appeared. Before 1970 the growth of the banking industry reflected the growth of the population and the increasing popularity of checking accounts. Since 1975, however, the growth (supported by mechanization) has reflected the increasing rate at which money flows from one investment to another.

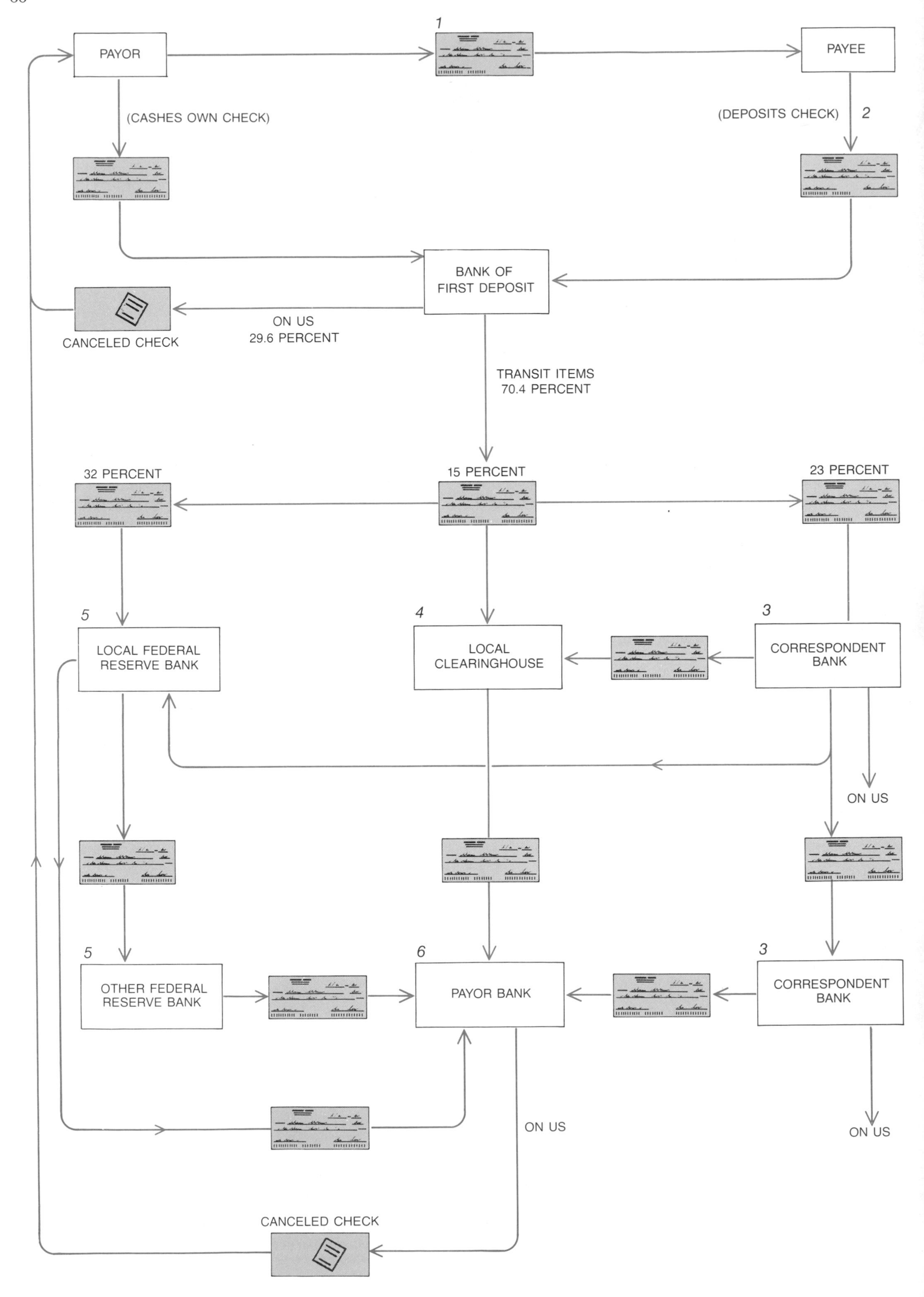

1
PAYOR
PAYEE
(CASHES OWN CHECK)
(DEPOSITS CHECK)
2
BANK OF FIRST DEPOSIT
CANCELED CHECK
ON US
29.6 PERCENT
TRANSIT ITEMS
70.4 PERCENT
32 PERCENT
15 PERCENT
23 PERCENT
5
4
3
LOCAL FEDERAL RESERVE BANK
LOCAL CLEARINGHOUSE
CORRESPONDENT BANK
ON US
5
6
3
OTHER FEDERAL RESERVE BANK
PAYOR BANK
CORRESPONDENT BANK
ON US
ON US
CANCELED CHECK

ible to the consumer: they are "bundled" into the total cost of bank and merchant services. Since most people are comfortable writing checks, many will be reluctant to change unless they get some benefit in the form of lower costs, greater convenience or extra service. In effect people may insist that they be rewarded for the labor they provide when they handle transactions electronically.

Service to individuals is only part of a bank's operations; relations with business organizations are at least as important. In this regard a major factor in mechanization has been the drive by businesses to improve their cash management. As interest rates have risen the number of ways in which a business can earn (or save) interest on short-term loans has grown. To earn or save the interest, however, the business must know its cash position accurately and be able to move its cash around quickly and economically. To meet these needs a variety of electronic terminal systems have evolved that enable businesses and other organizations to communicate directly with bank computers. By means of these terminals a business can keep track of its liquid assets almost minute by minute and can issue instructions for transferring them to where they can be best employed.

In order to move the funds rapidly banks in turn must be interconnected with one another by computer-to-computer telecommunications networks. The oldest and most important network is FedWire. FedWire, which began in 1918 when the Federal Reserve System leased a set of telegraph lines, serves to make settlements of the payments between banks that result from the totality of checks and other transactions individual banks have processed. The settlements are final, in that the interbank payments transacted by way of FedWire are guaranteed by the Federal Reserve System. Other major wire systems are CHIPS (Clearing House Interbank Payment System), which is operated by the New York Clearing House Association; BankWire, which is managed by an open consortium of U.S. banks, and S.W.I.F.T. (Society of Worldwide Interbank Financial Telecommunication), which originated as a European international system but now has many U.S. members. Among these, CHIPS provides the most dramatic example of growth and utilization.

CHIPS was established in 1970. It replaced an earlier system in which paper checks had been carried by messengers from major banks to the New York Clearing House, which provided a facility for settling local interbank accounts among its members. As the traffic in checks grew it became increasingly difficult for clerks in the individual banks to process the day's transactions with any assurance that outgoing payments (for which their bank would be responsible) were adequately covered by deposits and payments flowing in. The danger was either that the local checking system would lose its timeliness or that clerks would be forced to rely on their own judgment to decide whether or not to forward specific outgoing payments. In forwarding some payments the clerks rather than senior bank officers could be authorizing large amounts of credit.

The number of CHIPS transactions has increased by a factor of 20 in the first 10 years of the system's full operation. In terms of the dollar value handled its growth has been even greater, amounting to an average annual increase of some 40 percent. So rapidly has the flow of money into and out of business accounts increased that a typical member of CHIPS will process dollar values each day that can be tens of times the total worth of the bank itself. Meanwhile consumer-oriented automatic clearinghouses have been created to handle electronically operations such as the direct deposit of paychecks in an employee's bank account and the payment of periodic consumer bills for mortgages, rent and utilities. In the case of direct deposit, payroll data are transmitted to a clearinghouse, which rearranges the information so that each bank gets the individual payroll payments for the accounts it maintains. Direct deposit, encouraged strongly by the Federal Government for its own employees and for Social Security payments, is growing fairly rapidly; so far the consumer-payment side is clearly less popular.

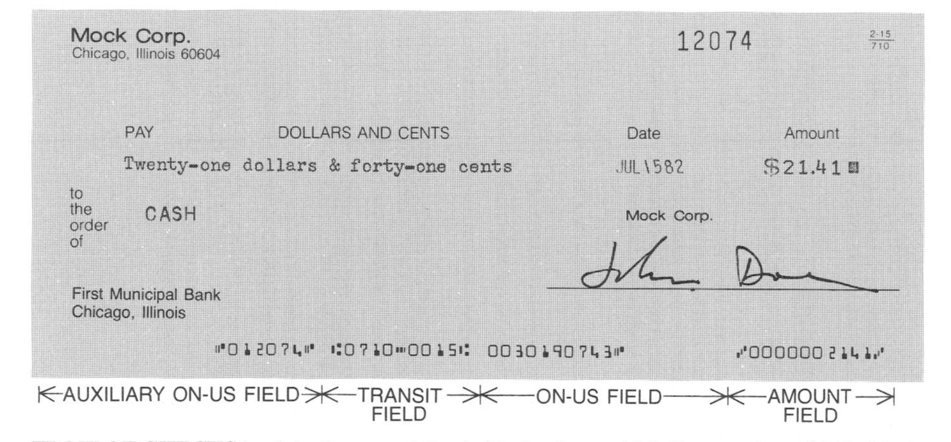

FLOW OF CHECKS back to the payor's bank (the bank on which they are drawn) is facilitated by mechanization that depends in turn on the machine-readable numbers across the bottom of the checks. The flow is diagrammed at the left. It begins with payments made by check (*1*). The amount of each such check is credited to the payee's account (*2*). Ultimately, however, the amount must be subtracted from the payor's account. Accordingly the check moves through a combination of three institutions: correspondent banks (*3*), which handle checking accounts for some banks and provide geographic coverage for others; clearinghouses (*4*), which receive the checks banks have credited to their own accounts and sort the checks for distribution to the banks on which they are drawn, and Federal Reserve Banks (*5*), which act as clearing houses. On the average a check in transit is processed by nearly two banks, including the payor's bank (*6*). The machine-readable numbers on checks are shown above. They are printed on checks in an ink containing an iron oxide, so that the numbers are magnetized and can be "read" by sorting machines and computer-input devices. (Increasingly the net transactions among banks are being settled electronically.) On-us fields are at the discretion of the payor's bank (or its correspondent): they identify the payor's account and may record the number of the check itself. The amount field indicates the dollar amount of the check as entered on the check by the payee's bank. The transit field guides the check through the banking system. Its first two digits specify which of the 12 Federal Reserve districts in the U.S. includes the payor's bank. Its third digit specifies a Federal Reserve office in that district (or a special arrangement for collection of funds). Its fourth digit specifies a state in the district (or a special collection arrangement). The remaining four digits identify the payor's bank specifically. Information in the transit field also appears (for manual sorting) at the upper right-hand corner.

Overshadowing all these changes has been the growth of bank-sponsored credit cards. Banks were relatively slow to offer credit cards in force. Individual banks had offered cards in the 1950's, and some regional systems evolved late in that decade, but the development of the two national bank systems, called Bank Americard and Master Charge at first, did not take place until the late 1960's. Growth after that was rapid, and the card systems have long since replaced many merchants' credit operations as well as cash and checks.

What has all this meant for banks, their customers and society in general? For one thing, it is clear that without mechanization many of the paper-based systems simply could not have coped with the growth in the use of bank services over the past two decades. Not only would labor costs have risen greatly but also, given the need for the careful checking and balancing of individual entries the banking industry requires, the sheer handling of paper documents would have made it impossible to keep the conduct of transactions timely. The result would no doubt have been some combination of higher prices, poorer service and limited growth. Instead productivity in the processing of checks has probably more than doubled between 1960 and 1980. The growth in productivity attributable to the wire networks has almost certainly been even larger.

In a subtle way mechanization has contributed to a change in the nature of banking. Because of Government regulations that restrict the activities of banks, the banks have lost ground to other financial organizations as intermediaries between those who have money available and those who want to borrow money. For example, large corporations now transact short-term loans with each other rather than conducting the equivalent transactions with banks. The loans are mediated by underwriting and securities companies. Moreover, money-market funds have replaced savings accounts for many individuals. Meanwhile, however, and partly as a result of high capacity, efficiency and low costs, the banks have increased their activity as movers of funds. In the process the operations of the financial intermediaries that now compete with banks have been supported and even facilitated.

Employment in banking has not suffered from mechanization. Because increases in productivity have been accompanied by even larger increases in the demand for bank services employment grew by 50 percent between 1970 and 1980, from slightly over a million to nearly 1.6 million. Much of the current flow of money through banks is a response to high inflation and volatile interest rates, and the flow may drop if a stabler economic environment returns. Regulatory reforms may remove the restrictions banks now face. The industry has nonetheless been restructured, and a complete return of banks to their traditional practices is unlikely.

The distribution of goods has always

UNIVERSAL PRODUCT CODE is the set of thick and thin bars printed on many supermarket items now sold in the U.S. It encodes 12 digits. Six of them, to the left of the "center guard pattern," are each represented by a light space, a dark bar, a second space and a second bar. The other six, to the right of the center-guard pattern, are each represented by a bar, a space, a bar and a space. The arrangement enables the computer at a grocery checkout counter to determine whether the code has been scanned backward by the sensor at the counter. (In that case the computer inverts the data.) The 12 digits have various meanings. The first decoded digit, which also appears in "human-readable form" at the left of the pattern of bars, is called the number-system character. A zero signifies a standard supermarket item. The next five decoded digits (in this case 12345) identify the manufacturer of the item. The five digits after that (67-890) identify the item itself. The last digit, which does not appear in human-readable form, serves to confirm that the other 11 encoded digits have been scanned and decoded correctly. It is the smallest number that yields a multiple of 10 when it is added to the sum of the second, fourth, sixth, eighth and 10th decoded digits plus three times the sum of the first, third, fifth, seventh, ninth and 11th digits. Corner markings define the area that should be blank around the set of bars. The set of bars was prepared by the Photographic Sciences Corp.

required financial services; in fact, many bank and insurance practices arose in response to the needs of those who transport goods. The mechanization of transport, however, clearly calls for more nonelectronic equipment than banking does, although transport too makes intensive use of electronic systems. The combination of nonelectronic and electronic equipment can be seen in the recent growth of containerized shipping. Containers themselves are an ancient form of technology, but their current level of service owes much to the availability of telecommunications and computers.

The modern container-ship era started in the late 1950's, when Malcolm McLean acquired the Pan Atlantic Steamship Company, a shipping organization that connected ports in the U.S. Northeast with ports on the Gulf of Mexico and in Puerto Rico. Having previously operated a trucking line, McLean decided to offer a "ferry service" for trucks, but one that left the truck wheels ashore. By having flatbed trailers carry containers overland and by lifting the loaded containers between ship and shore he speeded loading and discharge and could move goods in containers from their origin to their destination. A similar service between the U.S. West Coast and Hawaii was inaugurated by the Matson Navigation Company at about the same time. Both services flourished and others began to emulate them.

The subsequent progress of containerization was difficult. The standardization of container sizes and of the fittings with which cranes lift the containers was an early requirement. Regulation also influenced the pace: Puerto Rican and Hawaiian routes were domestic and relatively unregulated, but U.S. international shipping services were both highly regulated and highly subsidized by the Government. The subsidy mechanism tended to discourage the adoption of container technology; for example, it subsidized U.S. operators for crew costs higher than those of foreign operators but not for the costs of developing more fully mechanized ships and services.

Institutional barriers were also numerous. Shipping has many participants: shippers, freight forwarders (consolidators of cargo who make money from the differential in transportation rates between small quantities and larger ones), overland carriers, insurance companies, longshoremen, pier operators, agents and customs brokers (who are hired to expedite the formalities at the dockside) and the shipping lines themselves. There are likewise many regulators: the Federal Maritime Commission, the Maritime Administration, the Interstate Commerce Commission and the U.S. Customs Service. Each of these parties was affected differently by the introduction of containers, with some gaining advantages and others feeling threatened. Although the net financial effect of containerization was a considerable savings in costs, the distribution of the savings was worked out only slowly.

Containerization offers many benefits. Cargo is better protected from damage and pilferage and so insurance rates can be lower. Handling times and costs are greatly decreased; indeed, in many instances containers can move from their origin to their destination without being opened en route From the ship operator's point of view the major gain is the far greater utilization of his primary asset: his ships. With cargo-handling rates in port increased by a factor of 10 or more (from about 500 tons per day for conventional ships to 5,000 for container ships) container ships can spend far more time at sea earning money and less time in port, where all they accrue are costs.

To support container services a variety of telecommunications and computer systems are crucial. Since the ships are large, fast and spend little time in port, their cargo manifest is seldom complete when they are ready to leave. Thus manifests are best transmitted by telecommunications from computer to computer rather than being sent by air mail to destination ports. Even more important, it is necessary to keep track of the multitude of containers. Who owns each one? What type of container is it? Where is it? What does it carry, and from where to where? What is its condition? Does it need repairs? Who owes whom how much money for its use or maintenance? Where should it go next when it is empty? Is the supply of empties in a region adequate or inadequate? Where can empties be obtained? Small containerized shipping services can be operated without computers, but the large pools of containers necessary for efficient major operations would become hopelessly confused without computerized control.

The container ships themselves are increasingly mechanized. Typically the engine room in the newer ones goes unmanned on the night shift, because the instrumentation and control systems on the bridge are adequate for most operations. Crews have dwindled in size (although they remain above the minimum levels set by the U.S. Coast Guard), and because of the small amount of time the ships spend in port crews get long vacations and are given quarters more reminiscent of cruise ships than of traditional freighters. Matching the decrease in port time has been a decrease in the number of berths needed at a port. To be sure, considerable space must be available to marshal and store containers, but this space often is provided at new terminals on the outskirts of port cities. The inner-city waterfront previously devoted to shipping thereby becomes available for redevelopment as commercial, residential and recreational areas.

Since all ocean transportation requires some associated overland movement, containerization has also had an impact on trucks and railroads. Railroads in particular can take advantage of the large flows of cargo arriving on container ships by introducing efficient unit trains that move the containers to major inland destinations. Such services now operate coast to coast in the U.S., where they also are being utilized intensively to deliver bulk cargoes such as coal and ores. The cars making up bulk-cargo unit trains are coupled together for an extended period. The couplings allow each car to rotate about the long axis of the train, so that they can be unloaded by rotation without the labor-intensive operations of decoupling and recoupling, an example of a simple technology having quite impressive results.

New electronic equipment is being applied in all forms of transportation. Computer-based systems are beginning to provide brokerage services that bring together empty trucks and cargo awaiting movement. Microprocessors installed in vehicles now analyze data they get from sensors to improve the performance of all types of engines. They will come to have a role in other forms of vehicle control. In one potential application microprocessors would be given information on the "consist" of a train (the order and weight of each car in the train) and use it to guide the train's braking. The results would likely be lower fuel consumption and less wear and damage to both the train and its cargo. Up to now, however, the most widespread application of new electronic systems undoubtedly has been in air transportation.

Consider electronic reservation systems. The first was introduced in 1963 by American Airlines, Inc. It was devised in a joint effort with the International Business Machines Corporation. Before then the airlines employed massive paper-based systems to keep detailed passenger records and had electromechanical display boards in their larger reservation offices to show available seats on planes. The displays were hard to read from a distance; in fact, agents sometimes had to rely on binoculars. The correlation between paper records and electromechanical ones often was quite poor. Between 1960 and 1980 the number of passengers on U.S. airlines grew by a factor of five. The speed, accuracy, ease of use and cost efficiency of electronic reservation systems clearly facilitated this growth. Since air transportation was an expanding industry, employment in it also increased (by 15 percent between 1970 and 1980).

More than most activities, transporta-

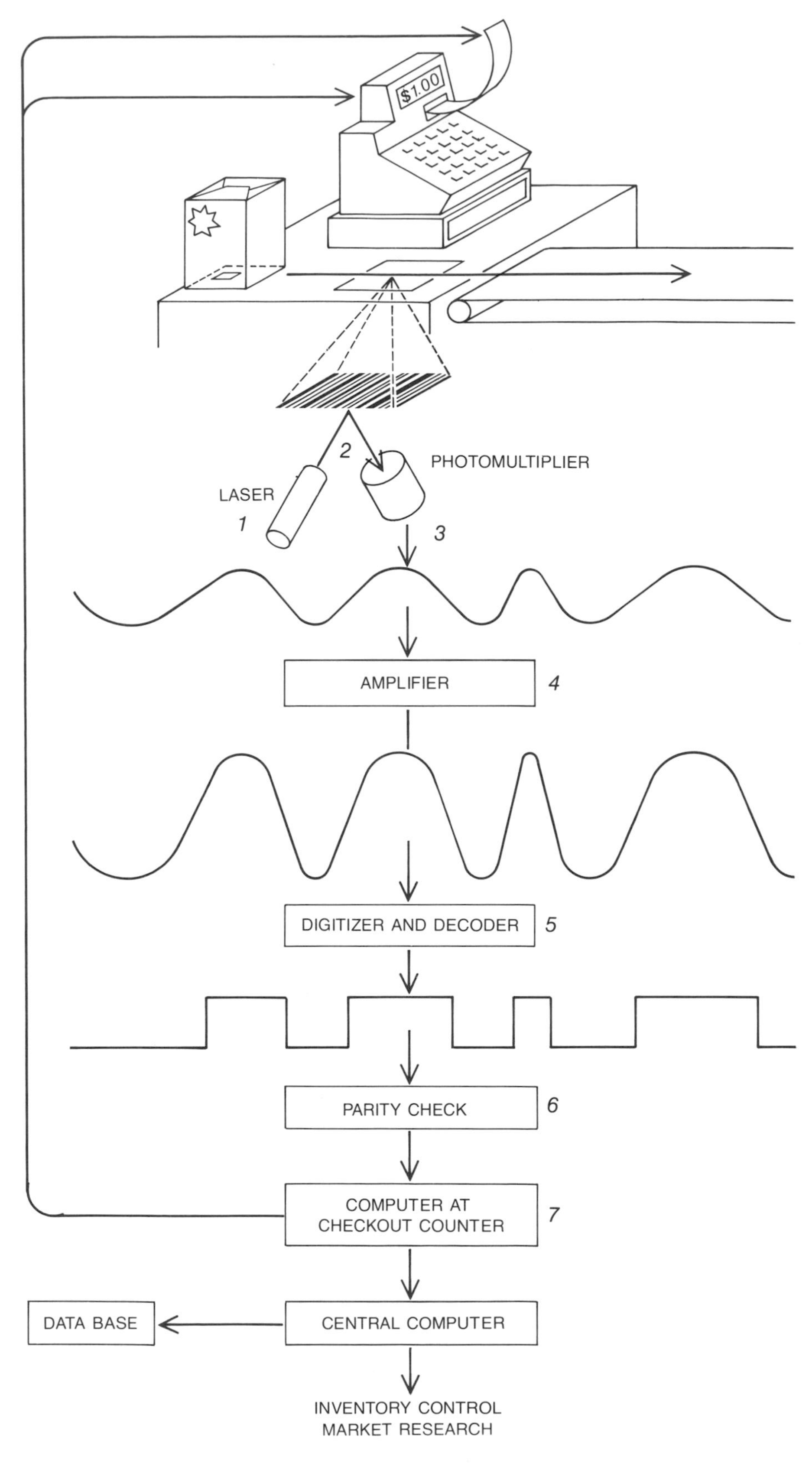

PRODUCT CODE IS SCANNED at the grocery checkout counter by a beam of laser light (*1*). The light reflected from the spaces between successive bars in the code (*2*) is converted into a continuous (analogue) electrical signal (*3*), which in turn is amplified (*4*) and "chopped" into a discontinuous (digital) signal (*5*). The latter signal is decoded, and the calculation of the 12th, or modulo-check, digit (*6*) confirms that the decoding is correct. Then a computer at the checkout counter (*7*) employs the decoded data to find in computer storage the price of the item. The decoded data (as input to a central computer) also guides inventory control, reordering, market research and the allocation of shelf space in stores. Bar codes of one type or another are now in service on library books, paperback books, magazines, order envelopes for photofinishing, ski-lift tickets, packages handled by delivery services and bags of blood in blood banks.

tion must be viewed as a system, since it often shunts goods from one mode of conveyance to another as it takes them from origin to destination. This suggests that the deregulation of transportation now under way in the U.S. will yield fewer and larger companies, with some of them operating in several modes. It seems likely that mechanization will accelerate the process, since the organizations best able to exploit the benefits of mechanization will usually be those with a degree of control over the entire system of moving goods rather than just one mode of it.

In each aspect of the final distribution of goods (that is, in wholesaling, warehousing and retailing) a variety of types of mechanization are in progress. An example in wholesaling is intercompany electronic-data interchange. Here the intent is to mechanize all aspects of order processing, including not only the transmission of orders to sellers but also the presentation to buyers, in electronic form, of current information on prices, discounts, special offers and the like. The technology is based on now-standard telecommunications networks connecting the computer terminals of the participants, but once again the critical requirement is for standardization.

The requirement arises because individual businesses usually have their own format for preparing price lists, orders and invoices. This creates no problems for people but presents major difficulties for computers. Usually an agreement on standards and formats is reached by trade associations. For example, the Food Marketing Institute, the Grocery Manufacturers of America, Inc., and four other trade organizations have coöperated in developing standards for transactions among distributors, brokers and manufacturers of food. These standards make up the Uniform Communications System (UCS) for the grocery industry. A slightly different approach is being taken for transportation documentation, an aspect of commerce that includes tariffs, manifests, control documents and arrangements for billing and payment. In this case the effort is being spearheaded by the Transportation Data Coordinating Committee, a nonprofit organization supported by dues from shippers, carriers and other members. The approach most often taken in all these efforts has been to standardize communication formats and protocols so that each participant is then able to develop computer programs that will translate between his private format and the communication standard.

In warehousing computers serve a wide variety of record-keeping and scheduling activities. A particularly valuable application is maintaining records of the location of items in storage. This makes possible random storage location, which can be almost twice as efficient as dedicated storage in utilizing the space in warehouses. Another major application is the mechanization of handling operations. The possibilities include aids to essentially manual operations by means of fork lifts and similar equipment that incorporate sensors and microprocessors; operator-controlled but nonetheless semiautomatic order-picking systems, and completely mechanized order-picking systems that can include automatic palletizers and depalletizers, devices that move cartons onto and off portable platforms. Fully automatic specialized warehouse systems were designed as early as 1958, but most of the efforts since then have been devoted to semiautomatic systems because such systems retain greater flexibility for responding to changes in the sizes and shapes of the objects being handled in the warehouse.

In retailing the focus of mechanization has been the cash register and other point-of-sale systems. The retailing of food offers the best example, largely because of the introduction of the Universal Product Code bars (UPC). Like the machine-readable numbers on checks, they are an instance of standardization for the sake of mechanizing the input of data for electronic processing. UPC bars are the set of thick and thin lines now printed on essentially all prepared-food items. At the checkout counter they enable the clerk to identify each purchase to an electronic terminal by passing the item over a photoelectric laser scanner built into the counter. The terminal then retrieves the price of the item from the store's central computer and prints it on a sales slip. At the same time the purchase record can be entered in an automatic inventory control and reordering system. That system in turn can produce data for market research, cost control in the store and shelf-space allocation. The least popular aspect of UPC is the display of prices. The merchant wants to post prices only on the shelves where items are stocked and save the labor cost of having them stamped on the items themselves. In some states, however, the merchant is being required by law to continue putting the price on each item. Nevertheless, the product-code systems are cost-effective, and after a slow start they are spreading fairly rapidly.

Other forms of mechanization in stores are quite common. Terminals have been developed to validate checks that a customer wants to cash by searching a data base to see if any of his previous checks has ever "bounced." Still other forms of mechanization are envisioned. Technically feasible, but not yet in significant service, is a terminal at the checkout counter that is connected to a local bank network. By inserting a plastic card known as a debit card the customer can directly debit his checking account by the amount of a purchase rather than paying for that purchase in cash or with a check.

The last aspect of commerce I shall touch on is telecommunications. It is an aspect of high importance. For one thing telecommunications is one of the fastest growing industries in the world. It has a range of recognized technical alternatives and potential applications that will take several decades to fully exploit. (The telephone network is so extensive that changes cannot come quickly. Imagine the effort that would be required to extend fiber-optic lines into every household and connect them to produce a visual-communication system.) At the same time the availability of adequate telecommunications is a prerequisite for many of the forms of mechanization described above.

Without a long history of incremental mechanization the telephone service we accept rather casually today would long since have become uneconomic or impossible to maintain. The most obvious episode in this history was the introduction of the dial system, which took the labor of establishing a telephonic connection between two parties and transferred it from the telephone operator to the caller. Over the years the dialing system has been expanded in geographic scope to both national and international operations. Moreover, the efficiency of long-distance operations has been enhanced by the installation of computer-based systems for recording calls and mechanizing the preparation of telephone bills. Within central telephone offices a series of technical advances has marked the transition from electromechanical switches to solid-state ones. The newer switches have lower maintenance requirements, are faster, have better transmission characteristics and more efficiently collect and maintain data for usage analysis and billing.

Many of these developments are based on sophisticated technology. The telephone system also illustrates, however, that major advances in productivity can be made with quite simple devices. Originally all telephone installations were "hard-wired," that is, the telephone lines were more or less permanently connected to the instruments. Then in the 1960's telephone companies introduced plug-in extension telephones. They nonetheless required that at least one hard-wired telephone be installed at each service number. More recently practice has changed further: the telephone wiring now put into houses has universal sockets that enable users to install their own telephones. Although this last change reduced pressure from manufacturers selling equipment designed to interconnect with telephone-company equipment, a major motive was the growing resistance of Government regulators to granting rate increases to cover escalating telephone-installation charges. Many of the telephone companies have now developed computer data bases that record the location of all the wiring and jack sockets in a home, so that people changing their residence can determine their needs in advance and arrange to pick up telephones at a telephone store and install the instruments themselves. A relatively simple set of changes in industry practice has thus allowed labor-sharing and a reduction in installation personnel.

Viewing the mechanization under way in commerce, one is reminded of some of the paintings of the elder Brueghel, with their almost frenetic energy and their display of the diversity and detail of human activities. Commerce is of course only one aspect of life and not the totality Brueghel viewed, and this narrower focus is largely responsible for the similarities I have noted that run through much of its mechanization.

First, although commercial activities provide most of our economic infrastructure, they also must rely on it. A technical innovation cannot be exploited widely if the infrastructure to support it is not available. To be sure, some institutions can take advantage of mechanization better than others, but the others cannot be left out or the process will fail to achieve the universality needed to make it effective. This is the reason the development of formal or informal standards is critical to widespread mechanization whether the standards are formats for data input and telecommunications or specify the size, shape or other characteristics of physical objects. The adoption of standards can trigger an entire set of steps toward mechanization. Still, the introduction of standards is not without risks. Standards introduce a rigidity that may be regretted later if technical advances open up better possibilities. Difficulties also can arise when different standards conflict. Banks have chosen magnetic technology for the input of data; retail stores have chosen optical technology. This may turn out to raise a barrier wherever a combined system is sought.

Second, mechanization calls for new institutional relations. Since commerce is a system and its benefits are distributed among many participants, agreement must be reached (or forced) on how the benefits of an innovation are to be shared. Otherwise there may be disincentives for any single participant to invest in the innovation; too few of the benefits may be gained by those who take the risks. The role of government in speeding, delaying or biasing mechanization, and therefore in the new institutional relations, is pronounced. In telecommunications and the financial industries much of the current deregulatory trend in the U.S. can be traced to

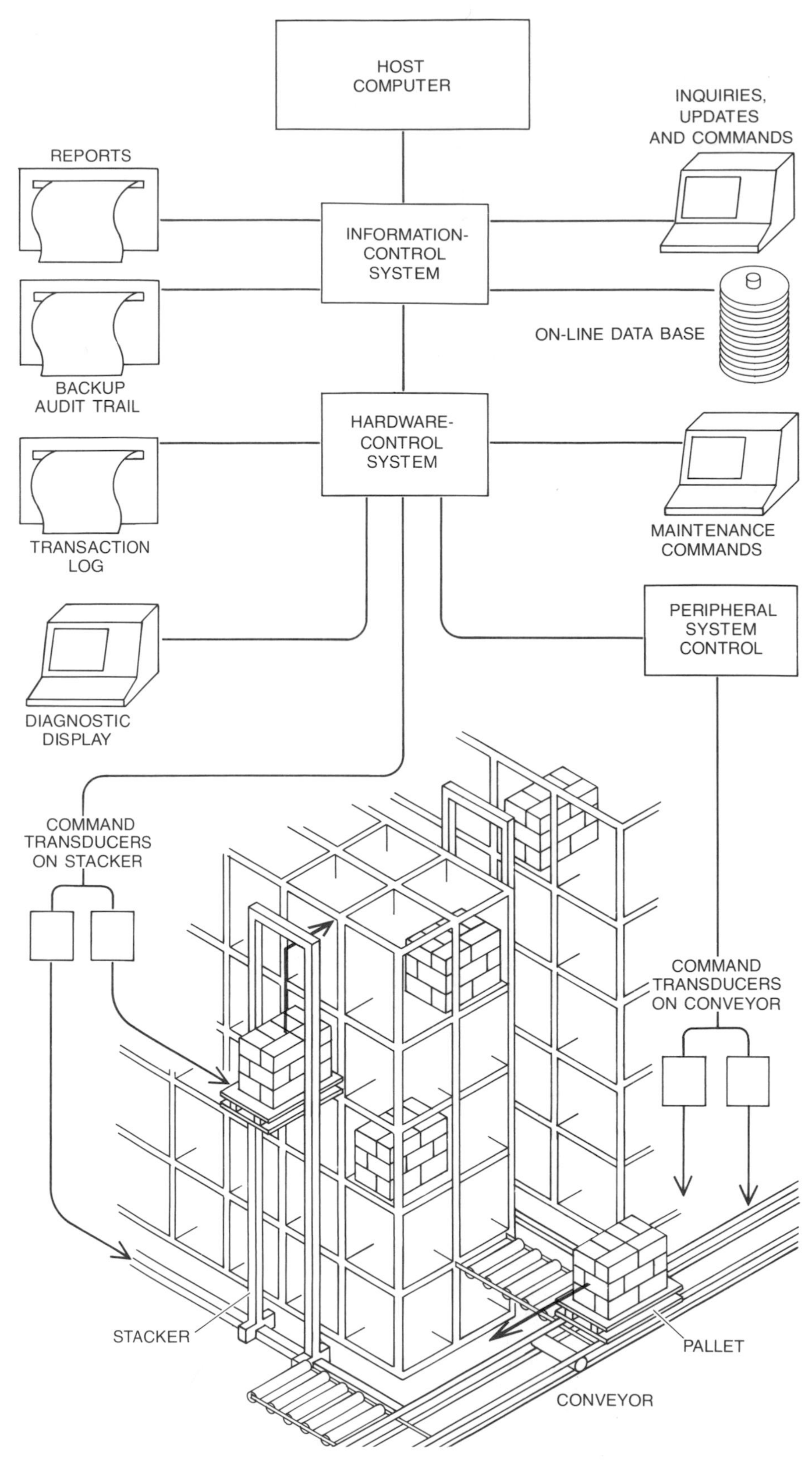

MECHANIZED WAREHOUSE IS CONTROLLED by two linked systems that are often in separate computers. One of them, the information-control system, calculates what might be termed a strategy for storage and retrieval. For example, it divides activity among the stackers, computes the shortest distance to empty cubicles and can select for removal the items that have been stored for the longest amount of time. In addition it continually updates its computerized record of what the warehouse holds. The other system is the hardware-control system. It governs the warehouse's moving equipment: its command signals are transduced into electrical currents that drive the motors on conveyors and stackers. Sensors on the conveyors and stackers in turn send signals to the hardware-control system so that the system can continually update a record of where the moving equipment is. The transaction log and the backup audit trail are printouts of what the systems have done. Often the computer systems that control a mechanized warehouse are linked electronically to a company's central (host) computer.

technical advances that either could not be exploited well in the regulated environment or could be used to bypass the intent of existing regulation. In the latter case the nation was being left with the rules but without the benefits the rules were supposed to encourage.

The benefits of the mechanization of commerce take many forms. Although labor productivity is an obvious example, the results of mechanization have almost always included advances in qualities such as capacity, speed, responsiveness, reliability and economy. These advances have sometimes helped to increase demand, so that displacement of labor has been avoided and employment has actually increased. In any case the level of activity and quality we now take for granted in many industries simply would not have been possible without mechanization.

So much for the benefits; what about the drawbacks? Some of them are related specifically to commerce; others belong to broader concerns about the impact of mechanization on society. In the first category the most publicized concern is the vulnerability of individuals that may result from mechanization in the various financial industries. This concern has already led to consumer-protection legislation covering the loss of credit cards, the actions required when errors in billing are found and similar matters. There remain, however, fears about fraud, theft and invasion of privacy.

There are a variety of measures to alleviate these fears. Some measures are legislative, some are technical. They will of course have a price: higher costs for financial institutions and their customers and more complex demands on consumers for identification during the entry of transactions. So far competition among financial institutions has discouraged individual efforts to implement the technical measures. The greatest current problem is the lack of a formal system for reporting and aggregating data on the frequency and scale of undesirable incidents. They will never be eliminated, but they can be controlled. One can anticipate a continuation of the perpetual war between the locksmith and the thief, now raised to a higher technological level that denies "employment" to the less skilled among white-collar thieves.

Another common concern is that the quality of life is being eroded by the depersonalization commonly associated with mechanization. Plainly the human touch is eliminated when one deals with a machine instead of a human being. Still, many of the examples of mechanization I have described are those where the human touch simply cannot be afforded, so that the real choice is between mechanized service or less service. Overall, therefore, human values may

MECHANIZED WAREHOUSE in Hatboro, Pa., is controlled by computers; it stores and retrieves pallets (portable platforms) that bear nonprescription pharmaceuticals (chiefly Formula 44 cough medicine and Lavoris mouthwash) manufactured by the Vick Chemical Company. When goods are being stored, the conveyor at the right in the photograph brings pallets (with their loads) to any one of seven aisles; there it transfers each pallet to an S/R (storage/retrieval) machine, which is also called a stacker. The stacker carries the pallet down the aisle, raises or lowers it and shuttles it into a storage cubicle. The stackers, 65 feet high, move among 14,000 cubicles; the aisles are 390 feet long. Each cubicle can bear a load of 2,500 pounds. The warehouse was constructed by Hartman Material Handling Systems, Inc.

be served better by accepting mechanization. In some situations, such as when services are needed by people who cannot speak the language of the country they are in, mechanized systems can be designed that are easier to deal with than a person who speaks only the native language. Furthermore, the younger generation quite obviously is not intimidated by the mechanized interface between consumers and electronic systems. The worst concerns probably arise for workers. Many mechanized systems (semiautomatic order-picking devices, for example) tend to isolate individual workers and break up normal social patterns.

More broadly, mechanized systems may actually result in overefficiency. Efficiency is almost always bought at a cost in flexibility and resilience. For example, the U.S. suffered from the Middle East oil crisis because its distribution system for petroleum was so efficient that inventories had been kept at a minimum. This offered little protection against an unanticipated interruption.

Finally, there are the fundamental concerns about displacement of labor. Who is to be displaced and how is the displacement to be handled? When the transition to a mechanized system is slow (as in containerization) or is in a growing industry (such as air transportation or banking), the direct impact is seldom severe; indeed, employment will often increase. When the industry is a mature one and the pace of change is rapid, the problem is not so easily solved. Efforts to retrain the displaced workers generally have a poor record even when the economy is growing rapidly. Most of the workers will feel a justifiable anguish at being severed from their accustomed livelihood. The least skilled workers will suffer disproportionately and are the least able to fend for themselves. This is a broad social problem whose solution will require educational reform and a new emphasis on encouraging the early development by young people of skills that are needed in an increasingly mechanized world. In this regard it is worth noting that future mechanization will probably broaden the range of those affected. Levels of middle management have already been eliminated by some forms of mechanization, and the spread of artificial intelligence may eventually affect even skilled professionals.

So far the demand for new services has been almost open-ended. At some point a fundamental restructuring of business and social life will no doubt come; at present it appears to be more than a decade away. The restructuring will be hard on many people, particularly those whose training and skills limit the types of employment they can seek. The immediate challenge, and a hard one, is to manage the restructuring in a more humane way than was typical of the original Industrial Revolution.

6

THE MECHANIZATION OF OFFICE WORK

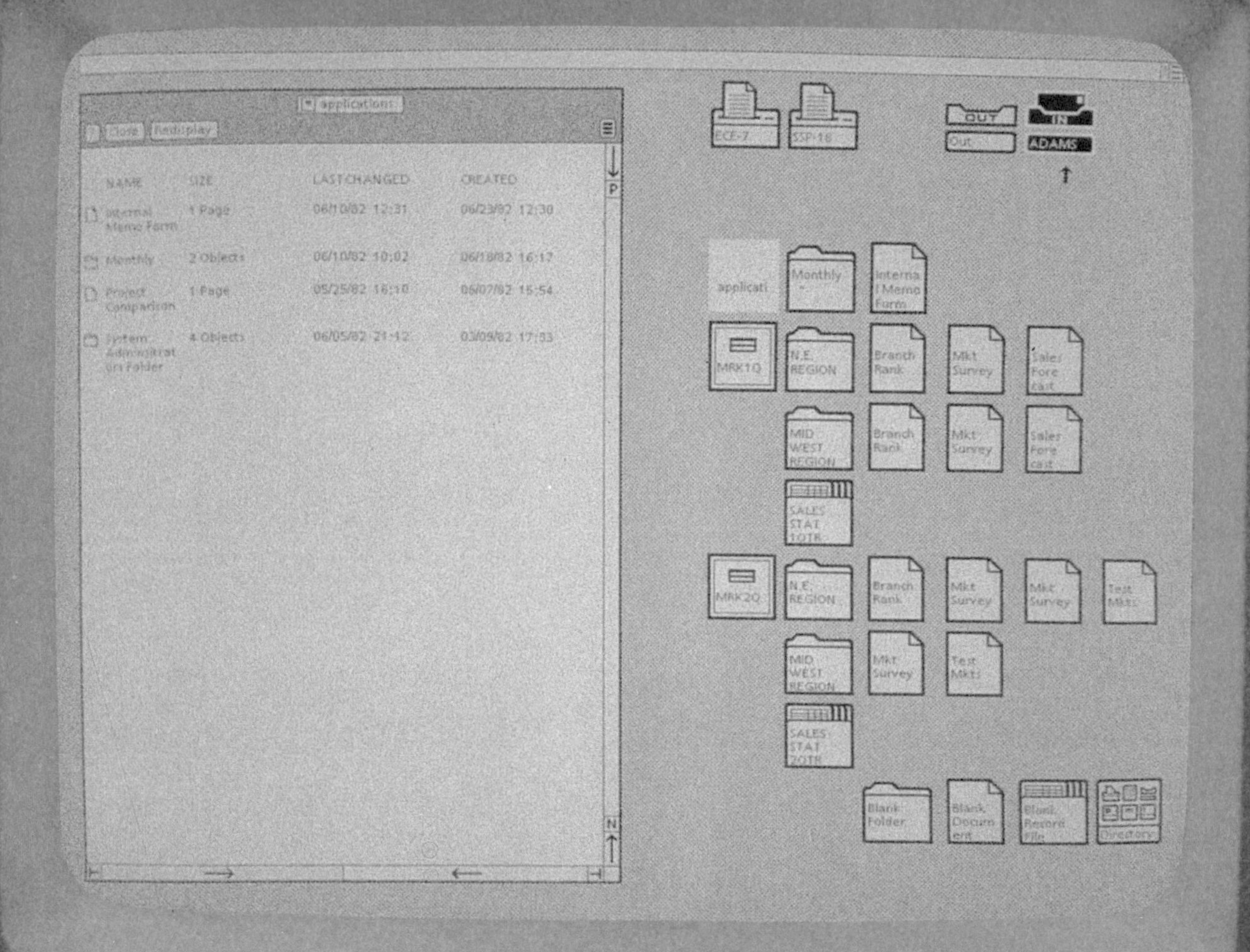
XEROX
applications
NAME
SIZE
LAST CHANGED
CREATED
Monthly
2 Objects
06/10/82 10:02
06/18/82 16:17
4 Objects
ECE-7
OUT
IN
ADAMS
applicati
Monthly
Internal Memo Form
MRK1Q
N.E. REGION
Branch Rank
Mkt Survey
Sales Fore cast
MID WEST REGION
SALES STAT 1QTR
MRK2Q
Test Mkt
SALES STAT 2QTR
Blank Folder
Blank Docum ent
Directory

The Mechanization of Office Work

by VINCENT E. GIULIANO

The office is the primary locus of information work, which is coming to dominate the U.S. economy. A shift from paperwork to electronics can improve productivity, service to customers and job satisfaction

Mechanization was applied first to the processing of tangible goods: crops in agriculture, raw materials in mining, industrial products in manufacturing. The kind of work that is benefiting most from new technology today, however, is above all the processing of an intangible commodity: information. As machines based mainly on the digital computer and other microelectronic devices become less expensive and more powerful, they are being introduced for gathering, storing, manipulating and communicating information. At the same time information-related activities are becoming ever more important in American society and the American economy; the majority of workers are already engaged in such activities, and the proportion of them is increasing. The changes can be expected to profoundly alter the nature of the primary locus of information work: the office.

An office is a place where people read, think, write and communicate; where proposals are considered and plans are made; where money is collected and spent; where businesses and other organizations are managed. The technology for doing all these things is changing with the accelerating introduction of new information-processing machines, programs for operating them and communications systems for interconnecting them. The transformation entails not only a shift from paper to electronics but also a fundamental change in the nature and organization of office work, in uses of information and communications and even in the meaning of the office as a particular place occupied during certain hours.

Office mechanization started in the second half of the 19th century. In 1850 the quill pen had not yet been fully replaced by the steel nib, and taking pen to paper was still the main technology of office work. By 1900 a number of mechanical devices had established a place in the office, notably Morse's telegraph, Bell's telephone, Edison's dictating machine and the typewriter.

In 1850 there were at most a few dozen "writing machines" in existence, and each of them was a unique, handmade creation. Typewriters were among the high-technology items of the era; they could be made in large numbers and at a reasonable cost only with the adoption and further development of the techniques of precision manufacturing with interchangeable parts developed by Colt and Remington for the production of pistols and rifles during the Civil War. By the late 1890's dozens of companies were manufacturing typewriters of diverse designs, with a variety of layouts for the keyboard and with ingenious mechanical arrangements. (Some even had the type arrayed on a moving, cylindrical element and thus were 70 years ahead of their time.) By 1900 more than 100,000 typewriters had been sold and more than 20,000 new machines were being built each year. As precision in the casting, machining and assembly of metal parts improved and the cost of these processes was lowered, typewriters became generally affordable in offices and homes. The evolution of typewriter usage was comparable to what is now taking place—in only about a decade—in the usage of office computers and small personal computers.

With the typewriter came an increase in the size of offices and in their number, in the number of people employed in them and in the variety of their jobs. There were also changes in the social structure of the office. For example, office work had remained a male occupation even after some women had been recruited into factories. (Consider the staffing of Scrooge's office in Charles Dickens' "A Christmas Carol.") Office mechanization was a force powerful enough to overcome a longstanding reluctance to have women work in a male environment. Large numbers of women were employed in offices as a direct result of the introduction of the typewriter [see "The Mechanization of Women's Work," by Joan Wallach Scott, page 87].

The first half of the 20th century saw a further refinement of existing office technologies and the introduction of a number of new ones. Among the developments were the teletypewriter, automatic telephone switching, ticker tape, the electric typewriter, duplicating machines and copiers, adding machines and calculators, tape recorders for dictation, offset printing presses small enough for office use and data-processing equipment operated with punched paper cards. The new devices were accompanied by a rapid expansion in the volume of office communications and in the number of people engaged in white-collar work.

The first computers in offices were

ELECTRONIC DESKTOP is emblematic of the shift from paper to electronics, the central element in the mechanization of office work. The desktop is displayed on the screen of the Xerox 8010 Star, a personal work station designed for business and professional workers. The Star by itself can serve as a small computer, a word processor and a generator of graphic material; when it is linked to other devices in a local-area network, the Star becomes an information system with access to an organization's electronic files, printers and interoffice and long-distance communications facilities. No special computer skills are needed to operate the work station. The screen shows a number of "icons" (*right*) representing familiar office objects, such as file drawers, file folders, individual documents, an "in" box and an "out" box. The worker sets up his own electronic desktop by manipulating the icons to store documents in folders and drawers, using a keyboard (not shown) and a "mouse" (*bottom*). The mouse is rolled about on the surface of the (nonelectronic) desk to control the position of a pointer on the screen. In the example shown a hypothetical sales manager named Adams has entered his name on the keyboard. His own desktop has been displayed, with a symbol showing there is material waiting in his in box. He has moved the pointer to the in-box icon, pressed the "select" button on the top of the mouse and pressed a key marked "open" on the keyboard, thereby calling up on the screen a list (*left*) of the contents of his in box. Now he can select, read and deal with any of the listed items. For example, he might call up the monthly report, revise it and have it printed.

crude and very expensive by today's standards. By the mid-1960's most large businesses had turned to computers to facilitate such routine "back office" tasks as storing payroll data and issuing checks, controlling inventory and monitoring the payment of bills. With advances in solid-state circuit components and then with microelectronics the computer became much smaller and cheaper. Remote terminals, consisting of either a teletypewriter or a keyboard and a video display, began to appear, generally tapping the central processing and storage facilities of a mainframe computer. There was steady improvement in the cost-effectiveness of data-processing equipment. All of this was reflected in a remarkable expansion of the computer industry. The late 1960's and the 1970's also saw the advent of inexpensive copiers, minicomputers, small and affordable private automated branch exchanges (electronic switchboards), the word processor (the typewriter's successor) and then, toward the end of the 1970's, the microcomputer.

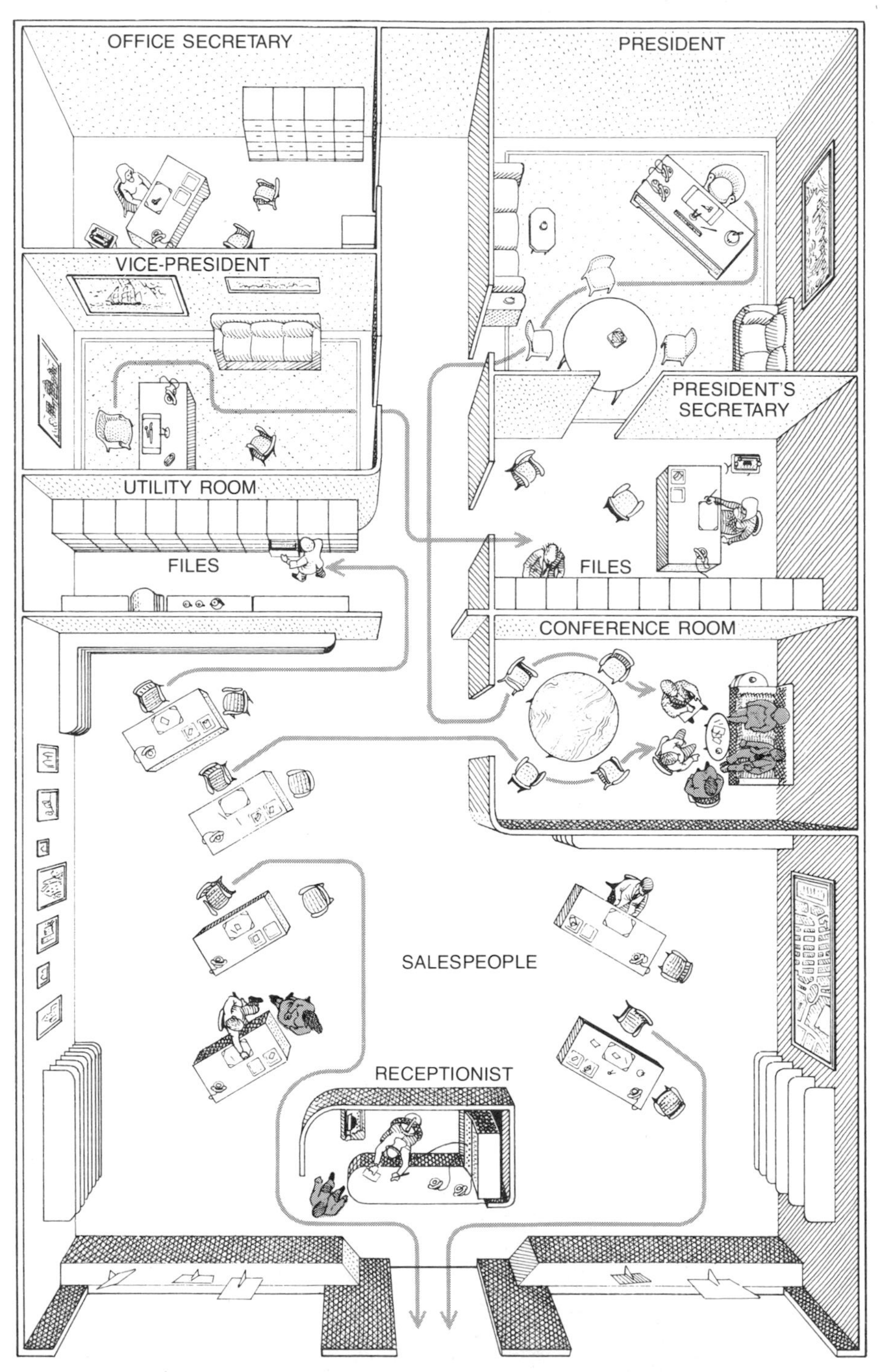

THREE STAGES OF OFFICE ORGANIZATION are defined by the author: preindustrial, industrial and information-age. Preindustrial organization dates back to the mid-19th century but is still typical of most professional, small-business and even corporate-management offices today. It is represented here by a hypothetical real-estate brokerage. There is little systematic organization. Each person does his job more or less independently, moving about as necessary (*gray lines*) to retrieve a file, to take a client to see a property or to attend a meeting where the sale of a house is made final (*color*). Individuals can have different styles of work, and human relations are important. The preindustrial model of office organization can still be effective for some small operations. Conversion to information-age methods is fairly easy.

An anthropologist visiting an office today would see much that he would have seen 25 years ago. He would see people reading, writing on paper, handling mail, talking with one another face to face and on the telephone, typing, operating calculators, dictating, filing and retrieving files from metal cabinets. He would observe some new behavior too. He would see a surprising number of people working with devices that have a typewriterlike keyboard but also have a video screen or an automatic printing element. In 1955 the odds were overwhelming that someone working at an alphabetic keyboard device was female and either a typist or a key-punch operator. No longer. The keyboard workers are both female and male and the typewriterlike devices now accomplish an astonishing variety of tasks.

Some of the keyboard workers are indeed secretaries preparing or correcting conventional correspondence on word processors. Other workers are at similar keyboards that serve as computer terminals. In one office they are managers checking the latest information on production performance, which is stored in a corporate data base in the company's mainframe computer. Economists are doing econometric modeling, perhaps calling on programs and data in a commercial service bureau across the continent. Librarians are working at terminals connected to a national network that merges the catalogues of thousands of participating libraries. Attorneys and law clerks are at terminals linked to a company whose files can be searched to retrieve the full text of court decisions made anywhere in the country. Airline personnel and travel agents make reservations at terminals of a nationwide network. Some of the devices are self-contained personal computers that engineers and scientists, business executives and many other people depend on for computation, data analysis, scheduling and other tasks.

Many of the users of terminals and small computers can communicate with one another and with their home offices through one of the half-dozen "electronic mail" networks now in existence in the U.S. A surprising number of people are doing these things not only in the office but also at home, on the factory floor and while traveling. This article was written with a portable personal computer at home, in a hotel in Puerto Rico and at a cottage in New Hampshire. I have drawn on information from personal files in my company's mainframe computer and have also checked

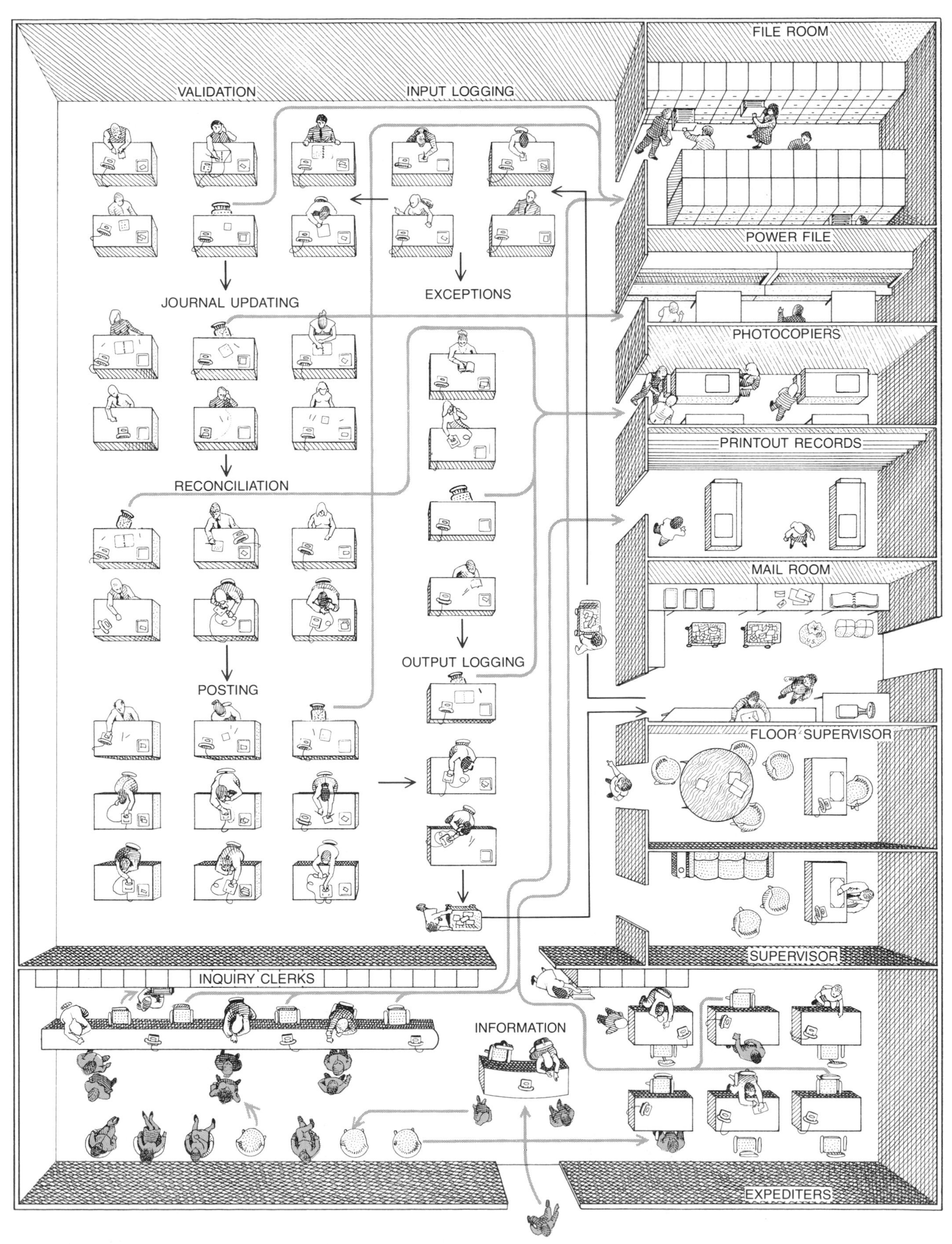

INDUSTRIAL OFFICE, essentially a production line, has been favored for operations handling a large number of transactions, as in this claims-adjustment department of an insurance company. Tasks are fragmented and standardized. Documents are carried from the mail room to the beginning of the production line and eventually emerge at the other end; the flow is indicated by the colored arrows. Successive groups of clerks carry out incremental steps in the processing of a claim; in general they leave their desks only to retrieve files or to examine computer printouts. If clients make inquiries, they are dealt with by clerks who may be able in time to answer a specific question but can seldom follow through to solve a problem. The work is usually dull. The flow of information is slow and service is poor.

parts of the text with colleagues by electronic mail.

What all of this adds up to is a shift from traditional ways of doing office work based mainly on paper to reliance on a variety of keyboard-and-display devices, or personal work stations. A work station may or may not have its own internal computer, but it is ultimately linked to a computer (or to several of them) and to data bases, communications systems and any of thousands of support services. Today the work stations in widest service handle written and numerical information. In less than a decade machines will be generally available that also handle color graphics and store and transmit voice messages, as the most advanced work stations do today.

My colleagues and I at Arthur D. Little, Inc., expect that by 1990 between 40 and 50 percent of all American workers will be making daily use of electronic-terminal equipment. Some 38 million terminal-based work stations of various kinds are by then likely to be installed in offices, factories and schools. There may be 34 million home terminals (although most of them may not function as full work stations). In addition we expect there will be at least seven million portable terminals re-

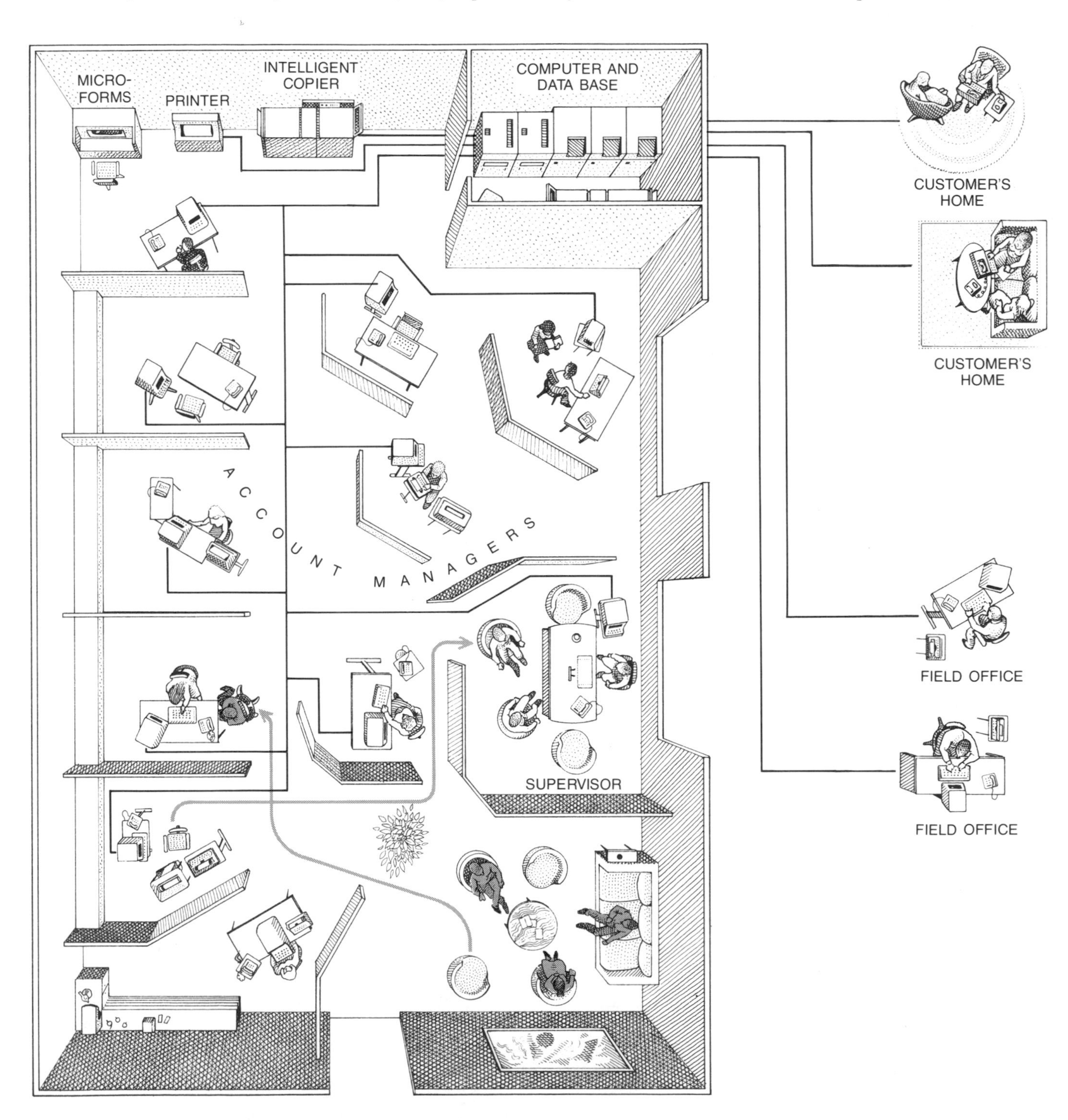

INFORMATION-AGE OFFICE exploits new technology to preserve the values of the preindustrial office while handling a large volume of complex information. The drawing shows an information-age claims-adjustment department. Each adjuster mans a work station, which is linked (*colored lines*) to a computer that maintains and continuously updates all client records. Each adjuster can therefore operate as an account manager, handling all operations for a few clients rather than one repetitive operation for a large number of clients. Necessary action can be taken immediately. Forms are updated and letters are written at the same work station that gives access to stored data, and the forms and letters can be printed automatically. The same facilities are available to adjusters visiting a client's home or working in one of the company's field offices (*right*). The work is more interesting, service to clients is improved and costs are reduced.

sembling today's hand-held calculators, most of them quite inexpensive.

Until recently most work stations and their supporting devices and data-base resources were designed to serve a single purpose: to prepare text, access stock-market data or make air-travel reservations, for example. The stockbroker's terminal started out as a replacement for the ticker tape, the word processor as a replacement for the typewriter. The first terminals therefore served as complete work stations only for people who were engaged in a more or less repetitive task.

Now the capabilities of the work station have been extended by developments in the technology of information processing, in communications and in enhancements of the "software," or programs, essential to the operation of any computer system. A variety of resources and functions have become accessible from a single work station. The stockbroker can not only check current prices with his terminal but also retrieve from his company's data base a customer's portfolio and retrieve from a distant data base information on stock-price trends over many years. Millions of current and historical news items can also be called up on the screen. He can issue orders to buy or sell stock, send messages to other brokers and generate charts and tables, which can then be incorporated into a newsletter addressed to customers. It is not only in large corporations that such tools are found. Low-cost personal computers and telecommunications-based services available to individuals make it possible for them to enjoy a highly mechanized work environment; indeed, many professionals and many office workers in small businesses have work-station resources superior to those in large corporations where the pace of office mechanization has been slow.

By the year 2000 there will surely be new technology for information handling, some of which cannot now be foreseen. What can be predicted is that more capable machinery will be available at lower cost. Already a personal computer the size of a briefcase has the power and information-storage capacity of a mainframe computer of 1955. For a small computer an approximate measure of performance is the "width" of the data path, that is, the number of bits, or binary digits, processed at a time. Computational speed can be represented roughly by the frequency in megahertz of the electronic clock that synchronizes all operations in the central processor. Memory capacity is expressed in bytes; a byte is a group of eight bits. The customary unit is the kilobyte, which is not 1,000 bytes but rather 2^{10}, or 1,024. Only three years ago a powerful personal computer had 48 kilobytes of working memory and an

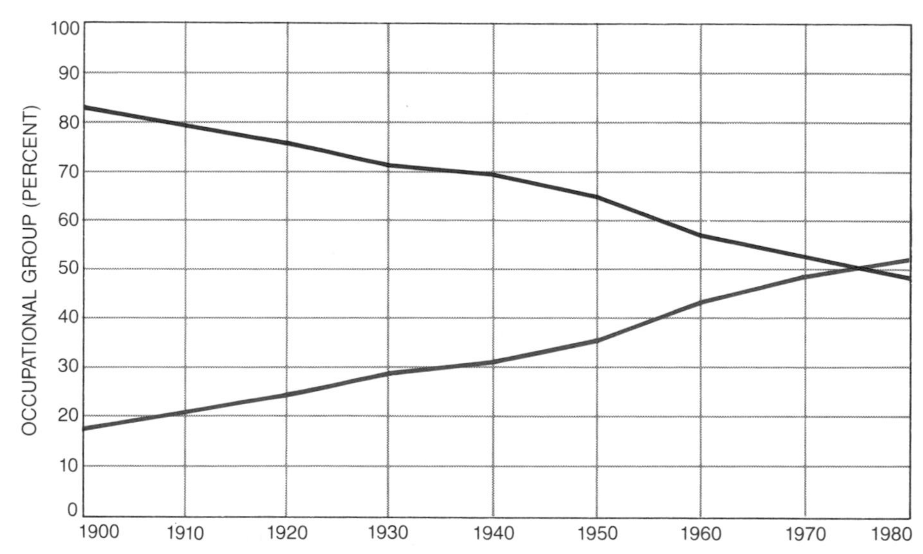

WHITE-COLLAR WORKERS now predominate in the U.S. economy. The curves show the percentage of the experienced labor force (from 1900 through 1950) and of all employed workers (from 1960 through 1980) that has been accounted for by workers in white-collar jobs (*colored curve*) and by blue-collar workers, service workers and farm workers (*black curve*).

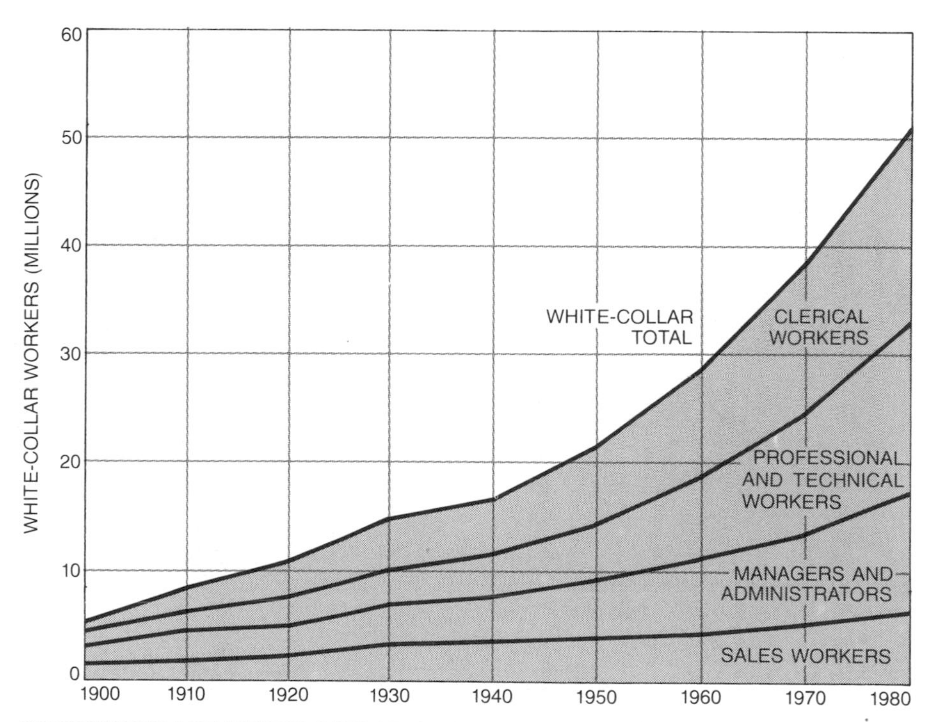

COMPOSITION OF WHITE-COLLAR GROUP has changed over the years. In 1900 clerical workers were the smallest category; now they are the largest. Most white-collar workers are office workers, and so office productivity has become a matter of increasing concern.

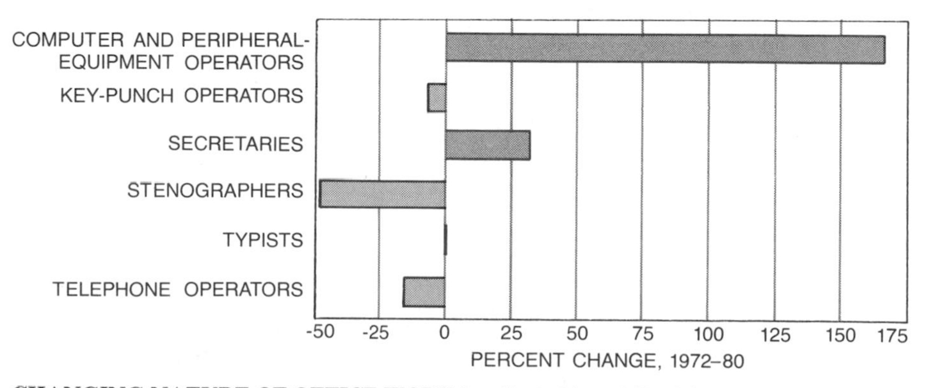

CHANGING NATURE OF OFFICE WORK is reflected in a shift of jobs within the clerical category. The bars show some of the changes from 1972 to 1980. Key-punch operators supply input for older computers. Telephone operators are being displaced by automatic switching.

eight-bit processor running at a rate of one megahertz.

Today about the same amount of money buys a machine with 256 kilobytes of working memory and a 16-bit processor chip that runs at four megahertz or more. Storage capacity and processing power will continue to increase—and their costs will continue to decrease—geometrically. By the year 2000 memory and processing power should be so cheap that they will no longer be limiting factors in the cost of information handling; they will be available as needed anywhere in an organization. The next 20 years will also see the continuing extension of high-capacity communications, of networks for the exchange of information between work stations and other computers and of centralized data banks. Together these developments will provide access to information, to processing capacity and to communications facilities no matter where the worker is or what time it is.

New technology inevitably affects the organization of work. One can define three evolutionary stages of office organization, which I shall designate preindustrial, industrial and information-age. Each stage is characterized not only by its technology but also by its style of management, personnel policies, hierarchy of supervisory and managerial staff, standards of performance and human relations among office workers and between the workers and their clients or customers.

The first two stages correspond to the well-understood artisan and industrial models of production; the nature of the third stage is only now becoming clear. The operation of a preindustrial office depends largely on the performance of individuals, without much benefit from either systematic work organization or machines. The industrial office organizes people to serve the needs of a rigid production system and its machines. The information-age office has the potential of combining systems and machines to the benefit of both individual workers and their clients.

Most small-business, professional, general-management and executive offices are still at the preindustrial stage. In a preindustrial office little conscious attention if any is paid to such things as a systematic flow of work, the efficiency or productivity of work methods or modern information technologies. What information-handling devices are present (telephones, copiers and even word processors) may be central to the operation, but there is no deliberate effort to get the maximum advantage from them. Good human relations often develop among the employees; loyalty, understanding and mutual respect have major roles in holding the organization together. An employee is expected to learn his job, to do what is wanted and needed and to ask for help when it is necessary. Varied personal styles of work shape the style of the operation and contribute to its success.

Preindustrial office organization generally works well only as long as the operation remains small in scale and fairly simple. It is inefficient for handling either a large volume of transactions or complex procedures requiring the coordination of a variety of data sources. If the work load increases in such an office, or if business conditions get more complex, the typical response is to ask people to work harder and then to hire more employees. Such steps are likely to be of only temporary benefit,

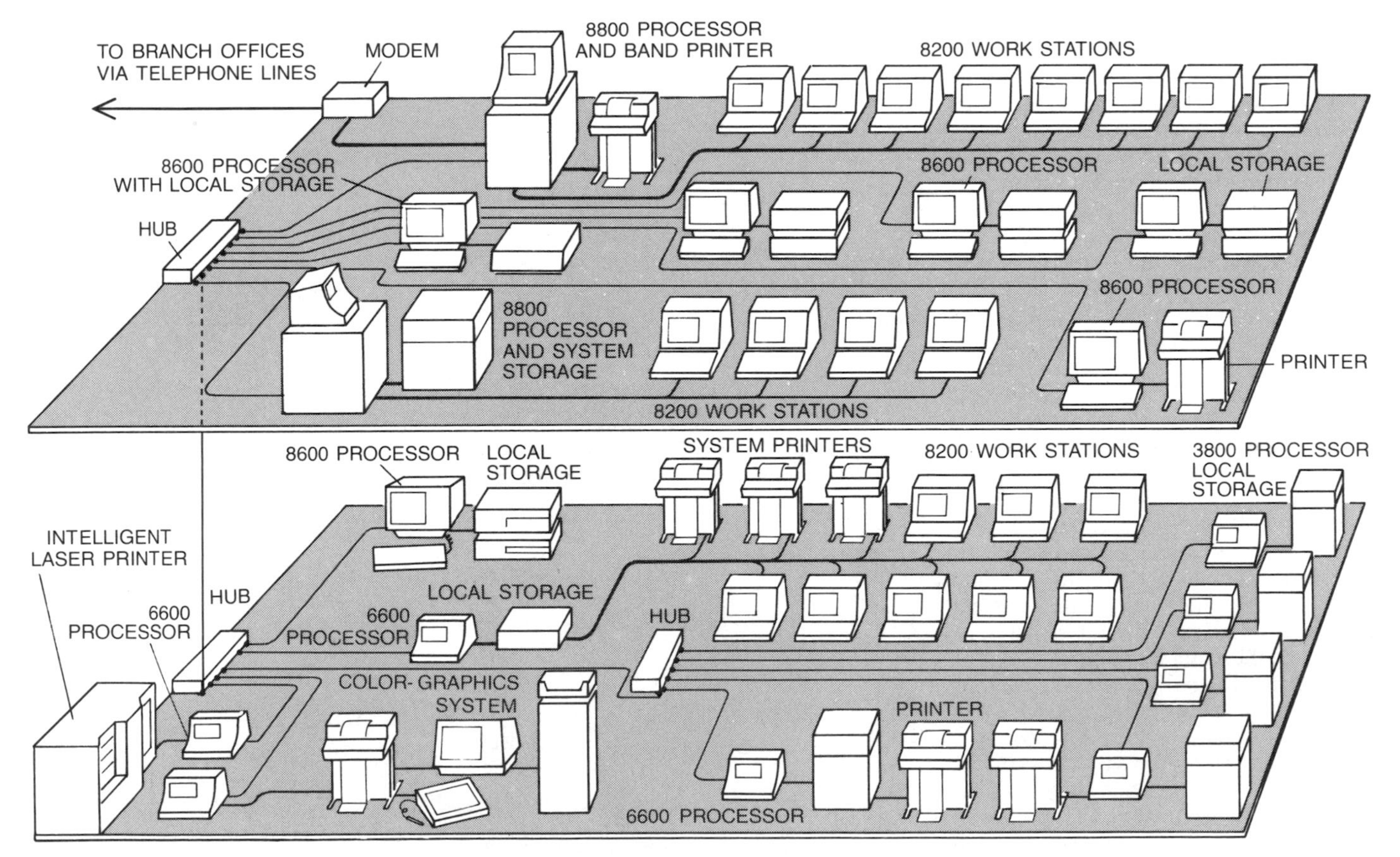

LOCAL-AREA NETWORK makes it possible for a large number of work stations in an organization to communicate with one another and to exploit the same data-storage and peripheral equipment. The system shown is the Datapoint Corporation's ARC (for "attached resource computer") network, in which as many as 225 processors (computers) can be linked by a system of coaxial cables (*colored lines*) and interfacing devices. Each processor in turn can be linked by wire (*black lines*) to a number of work stations, storage units or peripheral devices. Each processor has a "resource interface module" (RIM) by which it is connected to a cable leading to a "hub"; each input-output port (*black dots*) on a hub can be connected to a RIM or to another hub. Each RIM has an identification number and attends to any transmission addressed to that number. Traffic is controlled by a "token passing" scheme. Each processor controls the network in turn, taking over to transmit a brief packet of digital signals when it receives a token-passing message from the processor just ahead of it in line. In the system diagrammed here the resources shared by all the devices in the network include magnetic-disk storage units, printers, a generator of color graphics and a modem: a modulator-demodulator that converts digital signals into acoustic signals for transmission over telephone lines. There are also "local" disk units for storing material that is needed only by a single processor.

however. Without the help of additional systems or technology, effectiveness and morale may soon begin to break down.

One response to the limitations of preindustrial office organization has been to bring to bear in the office the principles of work simplification, specialization and time-and-motion efficiency articulated for factory work some 70 years ago by Frederick W. Taylor. The result is the industrial-stage office, which is essentially a production line. Work (in the form of paper documents or a folder of papers related to one customer) moves from desk to desk just as parts move from station to station along an assembly line. Each worker gets a sheaf of papers in an "in" box; his job is to perform one or two incremental steps in their processing and then to pass the paper through an "out" box to the next person, who performs the next steps. Jobs are simple, repetitive and unsatisfying. A worker may do no more than staple or file or copy, or perhaps check and confirm or correct one element of data. And of course everyone has to work together during the same hours in the same office to sustain the flow of paper.

The production-line approach has been considered particularly suitable for office activities in which the main job is handling a large volume of customer transactions, as in sending out bills or processing insurance claims. Many large production-line offices were instituted in the early days of computerization, when information had to be gathered into large batches before it could be processed by the computer; input to the machine then took the form of punched cards and output consisted of large books of printouts. Because early computers could do only a few steps of a complex process, the industrial office had to shape people's tasks to fit the needs of the machine. Computers and means of communicating with them have now been improved, but many large transaction-handling offices are still stuck at the industrial stage.

The industrial model of office organization is based on a deliberate endeavor to maximize efficiency and output. To create an assembly line the flow of work must be analyzed, discrete tasks must be isolated and work must be measured in some way. There is a need for standardization of jobs, transactions, technologies and even personal interactions. A fragmentation of responsibility goes hand in hand with bureaucratic organization and the proliferation of paperwork. Most of the workers have little sense of the overall task to which they are contributing their work, or of how the system functions as a whole.

The industrial office has serious disadvantages. Many errors tend to arise in a production-line process. Because of the subdivision of tasks efforts to cor-

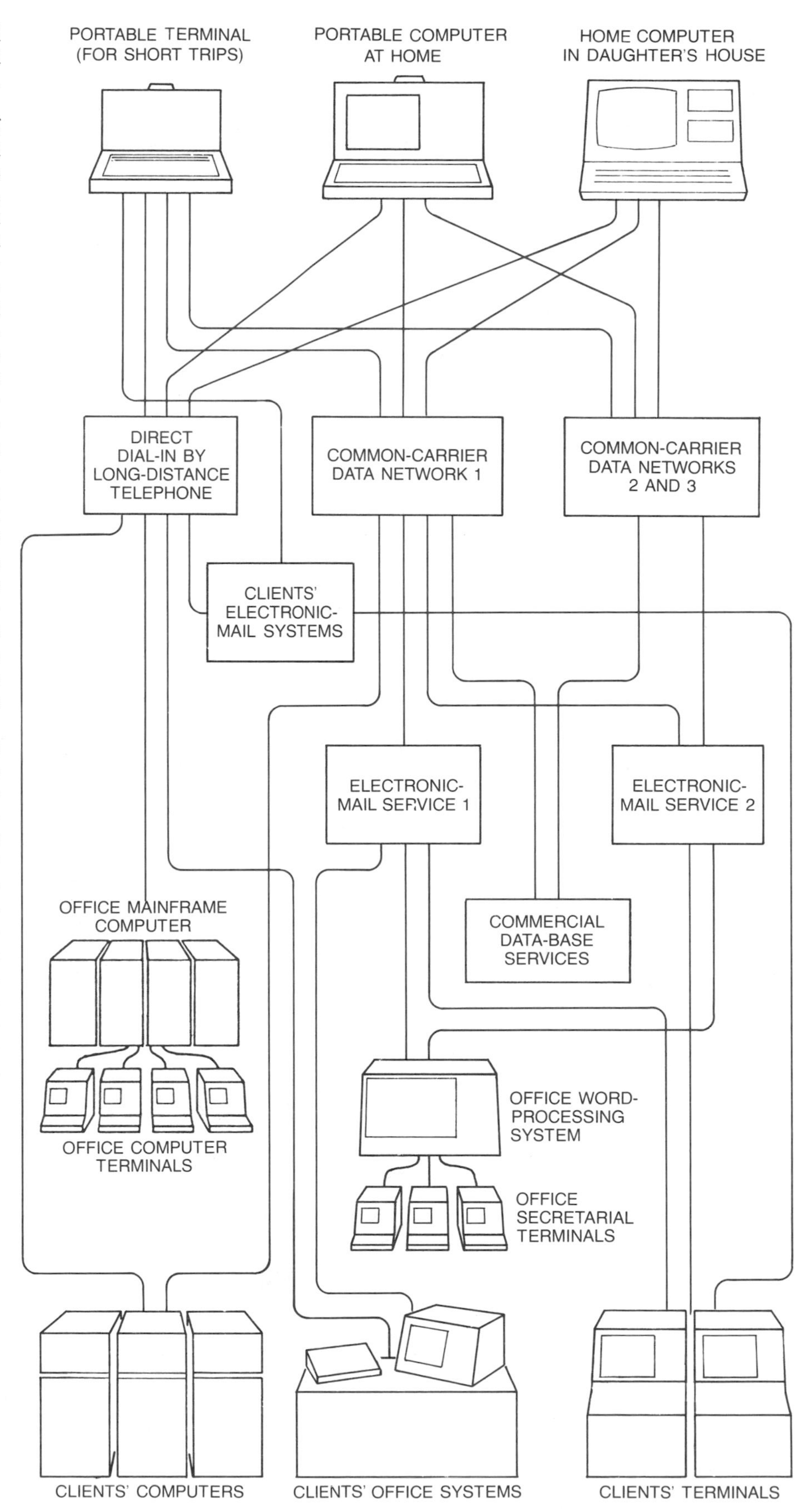

AUTHOR'S PERSONAL NETWORK enables him to work not only in his physical office but in a "virtual" office, which is to say almost anywhere and at any time. His own work station can be a portable terminal or either of two personal computers (*top*) or a work station in his physical office. Communication among physically separated elements of the network is by way of the public telephone system, with digital signals being converted into acoustic signals by portable acoustic couplers or modems. The author can write and edit material on any of the terminals and send it to his office or to clients. He can call up on the screen of a terminal any material stored in memory units at the office or in a commercial data base to which his company subscribes. He can also send and receive messages through two electronic-mail services.

rect errors must often be made without access to all pertinent information, with the result that the errors are sometimes not corrected but compounded. Moreover, production-line operations can be surprisingly labor-intensive and costly. As more people are hired to cope with an error rate that increases faster than the volume of transactions, the cost per transaction increases and efficiency declines.

Effective people do not want to stay in boring jobs; people who do stay often lack interest in their work, which becomes apparent to the customer. Even if workers do their best, the system may defeat them, and customer service is likely to be poor. Because a given item can take weeks to flow through the pipeline it is often difficult to answer customer inquiries about the current status of an account and even harder to take corrective action quickly. For example, a clerk may be able to check a sales slip and agree that a customer's bill is incorrect; in many instances, however, the clerk is able to change the account only by feeding a new input into the production line, with little assurance it will have the desired effect. As a result the billing error can be adjusted incorrectly or can be repeated for several months.

In the mid-1970's the recognition of these limitations, combined with the availability of new work-station information systems, motivated a few progressive banks and other service organizations with a heavy load of transactions to take the next step: they converted certain departments to a mode of operation more appropriate to the information age. The information-age office exploits new technology to preserve the best aspects of the earlier stages and avoid their failings. At its best it combines terminal-based work stations, a continuously updated data base and communications to attain high efficiency along with a return to people-centered work rather than machine-centered work. In the information-age office the machine is paced to the needs and abilities of the person who works with it. Instead of executing a small number of steps repetitively for a large number of accounts, one individual handles all customer-related activities for a much smaller number of accounts. Each worker has a terminal linked to a computer that maintains a data base of all customer-related records, which are updated as information is entered into the system. The worker becomes an account manager, works directly with the customer and is fully accountable to the customer.

Information is added incrementally to the master data base. The stored data are under the control of the worker, who can therefore be made responsible for correcting any errors that arise as well as for handling all transactions. Since information is updated as it becomes available there is no such thing as "work in process," with its attendant uncertainties. An inquiry or a change in status can be handled immediately over the telephone: the sales slip can be inspected, the customer's account can be adjusted and the bill that is about to be mailed can be corrected accordingly.

The design of effective systems and the measurement of productivity are

COMPUTER TERMINALS have a conspicuous place in an office of the Prudential Insurance Company of America in Parsippany, N.J., where claims are processed. Personnel who have identified themselves by entering a password at the keyboard of a terminal can retrieve information on an insured person's policy and claim, modify the information as necessary and add new information to the file.

still important in the information-age office with a large volume of transactions, but the context is different from that of the industrial office. Productivity is no longer measured by hours of work or number of items processed; it is judged by how well customers are served. Are they satisfied? Are they willing to bring their business back? Are they willing to pay a premium for a high level of service?

To the extent that the answers are yes the company gains an important competitive advantage. Even if cost cutting is not the only objective, the company can expect dramatic savings in personnel costs. Staff reductions of as much as 50 percent have been common in departments making the changeover to a work-station system. Those employees who remain benefit from a marked improvement in the quality of their working life.

The benefits of the information-age office are not limited to the transaction-intensive office. A similar transformation can enhance productivity, effectiveness and job satisfaction in offices concerned with management, general administration and research. Most such offices are still in the preindustrial stage. They can be transformed to the information-age stage by the introduction of such person-centered technologies as the work station and electronic mail.

PORTABLE TERMINAL is used by Malcolm Moran, a sports reporter for *The New York Times*, to cover a game at Shea Stadium between the New York Mets and the Montreal Expos. The terminal, a Portabubble 81 made by the Teleram Communications Corporation, is carried on out-of-town assignments by most *Times* reporters. Its magnetic-bubble memory holds between 9,000 and 20,000 words. The reporter can keep notes in the terminal's memory and write part of a story and store it for later transmission; he can have background material transmitted to him from the *Times*. Ordinarily the sports reporter writes his article at the end of the game and then transmits it to the computers at the *Times*. Sending and receiving is by means of an acoustic coupler. To file his story the reporter dials a telephone number and gets a go-ahead signal. Then he puts the telephone handset on the coupler and presses a button; the coupler converts the terminal's digital signals into acoustic signals and the story is transmitted at a rate of 300 words per minute. From the *Times* computers the article can be called up on terminals in the newsroom for editing and sent to the composing room for electronic typesetting.

Once most of the activities of a job are centered on the work station the nature of the office can be transformed in still another way: there is no longer any need to assemble all workers at the same place and time. Portable terminals and computers, equipped with appropriate software and facilities for communication (including the telephone), create a "virtual" office, which is essentially anywhere the worker happens to be: at home, visiting a client or customer, in a hotel or even in an airplane. The remote work station can communicate electronically with the central office and so it extends the range of places where written and numerical material can be generated, stored, retrieved, manipulated or communicated.

The effects of small-computer technology on the locale of work are analogous to those of the telephone. Because of the almost universal distribution of telephones it is not necessary to go to the office to call a customer or a co-worker, but until now it has been necessary to go there to write or dictate a letter, to read mail or to find something in a file. Now the work stations and ancillary electronic devices of an automated office can be linked to external terminals and personal computers. The job is no longer tied to the flow of paper across a designated desk; it is tied to the worker himself. The individual can therefore organize his own time and decide where and when he wants to do his work. Individuals who work best early in the morning or late at night can do so. A project team I have been working with for about a year has members in several East Coast and West Coast cities and rural areas, and we communicate regularly by electronic mail. The cost of the correspondence is about a tenth of the cost of regular mail per item, and it turns out that about half of the messages are generated outside of offices and outside of conventional working hours.

What will happen to the physical office? It has its virtues, after all. The office provides a home for organizations, a place for people to come together face to face and a work-oriented environment away from home. Many people need the structure of an office schedule; they like (or at least they are accustomed to) compartmentalization of the day and the week into time for work and time for other activities. Another role for the office is to house centralized forms of communications technology, such as facilities for video conferences, that are too expensive for the home. For these reasons and others I think the physical office will remain a part of working life, at least for as long as I am working. There will be continuing change, however, in how often some workers go to the office and in why they go there.

Many powerful factors are operating together to propel the transformation of office work. A complex set of feedback loops links economic and social change, new developments in information technology, the widespread adoption of the technology and the introduction of the new office organization the technology makes possible. The large number of information workers, for example, stimulates interest in enhancing their productivity. The concern for productivity serves to increase demand for technologies that can reduce the cost of handling information. Thus several trends reinforce one another to generate an ever stronger market for information products and services. The infiltration of the new devices into the workplace in turn creates an environment in which working electronically is the normal expectation of the worker.

Economics is a major factor. It is becoming far cheaper to communicate electronically than it is to communicate on paper. The transition to word processing from multidraft secretarial typing can reduce secretarial costs from more than $7 per letter to less than $2. Even more dramatic savings are associated with electronic mail, which can bring the cost of sending a message down to 30 cents or less. Electronic filing, in which a "document" is stored and indexed in a computer memory, brings further savings. (The highest-cost activities in manual correspondence are making multiple copies, filing them and retrieving them.) Such obvious reductions in cost are overshadowed by the savings in the time of managers and executives, the largest element by far in the cost of running an office.

The savings are becoming more significant each year as the cost of the electronic technology is reduced. For example, fast semiconductor memory is a tenth as expensive now as it was in 1975; the cost will drop by another factor of 10 by 1995. The result has been to bring into the individual consumer's price range information-handling capabilities that only a few years ago called for very expensive equipment.

As the market for mechanized work stations expands, more money is invested in research and development for communications, electronics, software, office-mechanization systems and the like. The time span between the development, introduction and obsolescence of a product becomes shorter. Each year brings a new generation of semiconductor devices; each generation makes possible a new set of applications. The dramatic improvement in products in turn builds demand for them and strengthens the trend toward office mechanization.

Whether a company's business is in farming, mining, manufacturing, transportation or retailing, its management, marketing, distribution and other operating controls are basically office-centered, information-handling activities. As the number of blue-collar workers decreases, the proportion of white-collar workers even in manufacturing organizations continues to increase. In virtually all commercial enterprises one finds executives, managers, clerks and secretaries; in most organizations there are also more specialized information workers, such as engineers and scientists, attorneys, salesmen, librarians, computer programmers and word processors. These people constitute the human-capital resources that can make an information-intensive economy viable.

Yet a tendency to think of white-collar workers in offices as support personnel, outside the economic mainstream, has tended to inhibit the transformation of office work. Physical activities that produce food, minerals and manufactured goods have been regarded as the only truly productive ones, whereas the handling of information has been considered necessary but essentially nonproductive. This way of looking at things (which may have been appropriate in an industrial society) persists today, even in the minds of economists who call for the "reindustrialization of America." It deeply affects the thinking of corporate management.

Even though most work in American society is information work and most such work is done in offices, the benefits of an increase in the productivity of office workers are not always within the field of view of managers. For those who retain a preindustrial view of office organization the very concept of productivity seems irrelevant or inappropriate in the context of offices or information work. Those who have an industrial-office orientation tend to focus on laborsaving measurements; the installation of new technology and a system for exploiting it is evaluated only in the context of cutting visible office costs.

It is in offices that the basic decisions are made that determine the cost-effectiveness of an entire organization. The office is the place where the timeliness of a decision or of a response can have immense consequences. If the office is ineffective, the organization must be ineffective. As it happens, moreover, a high degree of mechanization of the kind described in this article is much less expensive in the office than analogous mechanization is in the factory or on the farm.

The mechanization of office work is an essential element of the transformation of American society to one in which information work is the chief economic activity. If new information technology is properly employed, it can enable organizations to attain the following objectives: a reduction of information "float," that is, a decrease in the delay and uncertainty occasioned by the inaccessibility of information that is being typed, is in the mail, has been misfiled or is simply in an office that is closed for the weekend; the elimination of redundant work and unnecessary tasks such as retyping and laborious manual filing and retrieval; better utilization of human resources for tasks that require judgment, initiative and rapid communication; faster, better decision making that takes into account multiple, complex factors, and full exploitation of the virtual office through expansion of the workplace in space and time.

MAILMOBILE, a driverless battery-powered delivery vehicle made by Bell & Howell, mechanizes intraoffice deliveries. Here it is negotiating a curve as it makes its way through the research department of Merrill Lynch and Company. The vehicle follows a chemical pathway, which is easily applied and modified to trace any route from the mail room through the office and back to the mail room. An emitter of ultraviolet radiation under the vehicle makes the chemical fluoresce; an optical sensor detects the fluorescent path. The Mailmobile moves at about one and a half feet per second (one mile per hour), beeping and flashing blue headlights. It stops at pickup and delivery sites designated by a coded pattern in the chemical pathway. Bumpers stop the vehicle on contact with a person or another obstacle. An "intelligent" version is being introduced that can be directed to choose among alternate paths or to board an elevator.

7

THE MECHANIZATION OF WOMEN'S WORK

The Mechanization of Women's Work

by JOAN WALLACH SCOTT

When it began two centuries ago, it was characterized by low pay and occupational segregation. The same holds true today, although women are entering the labor force in larger numbers

It is frequently assumed that the mechanization of work has a revolutionary effect on the lives of the people who operate the new machines and on the society into which the machines have been introduced. For example, it has been suggested that the employment of women in industry took them out of the household, their traditional sphere, and fundamentally altered their position in society. Both advocates and critics of mechanization have shared the assumption. As women began to enter factories in increasing numbers in the 19th century Jules Simon, a French politician, warned that "a woman who becomes a worker is no longer a woman." Friedrich Engels, on the other hand, thought women would be liberated from the "social, legal and economic subordination" of the family by technological developments that made possible the recruitment of "the whole female sex... into public industry." Thus two observers could have diametrically opposed views on the value of mechanization for women without either one doubting that mechanization would transform women's lives.

Simon and Engels and many others imputed such transforming power to technology partly because they thought the capacity of mechanization to alter human relations was inherent in the machines themselves and hence capable of powerfully affecting the social context in which the machinery was utilized. This hypothesis has now been seriously questioned by historians, particularly those investigating the history of women. Scholars who have examined the experience of women in industrial society have concluded that such innovations as the spinning jenny, the sewing machine, the typewriter, the telephone, the vacuum cleaner and the computer have not fundamentally changed the economic position of women or the prevailing evaluation of women's work. Dramatic technological changes did not result in equally dramatic social changes. For example, the employment of young women in early textile mills was often an extension of an older pattern of employment of young single women. The employment of women in offices was the result of the separation of secretarial work from administrative work and the consequent creation of a class of jobs with little opportunity for advancement; the new jobs were often thought to constitute "women's work." The increased employment of married women in the 20th century, which was a substantial social change, had less to do with mechanization than it did with other economic and demographic trends.

It is undeniable that some aspects of women's work have changed considerably in the past 200 years. Work has moved from the household to the office or factory; in many cases it has become white-collar work rather than blue-collar work. In certain essential respects, however, the work that women do has changed little since before the Industrial Revolution. Occupations are still frequently segregated according to sex. Women as a group are paid less than men. Their work in many cases calls for a relatively low level of skill and offers little opportunity for advancement. For women with families household labor remains demanding, even if they can afford household appliances their grandmothers would have found miraculous. A decade of historical investigation has led to a major revision of the notion that technology is inherently revolutionary, at least as the notion applies to women. The available evidence suggests that on the contrary mechanization has served to reinforce the traditional position of women both in the labor market and in the home.

Mechanization has, of course, had a revolutionary effect on the processes by which goods are made and the organization of the workers who make them. Steam-driven spinning and weaving machines introduced in the 19th century could make in minutes as much thread or cloth as hundreds of individual artisans had been able to make in days or weeks. The new machines simplified the tasks required to make finished goods, divided the work into small, repeated operations and brought together under one roof large numbers of people doing similar work.

Many of those who went to work in the new mills in Europe and the U.S. were young women who had left farms and spinning wheels to take jobs as machine operatives. The contrast between the experience of the farm girl and that of the mill worker was dramatic, and it could be overwhelming. The contrast is evident in a thinly fictionalized account written by a woman who had been a new arrival at a textile mill in Lowell, Mass.,

SPINNING YARN, traditionally done by women at home, moved into factories as the result of the mechanization of the textile industry. The photograph on the opposite page shows the spinning room of Pacific Mills in Lawrence, Mass., in 1915. Spinning is the drawing out and twisting of cotton fibers to form yarn. It was originally done by hand with the distaff and the spindle. The first mechanical advance was the invention of the spinning wheel in the Middle Ages. Before the establishment of textile mills most spinning was done at home with spinning wheels. The development of power looms made it impossible for women to supply enough yarn by this means. To increase the supply several innovations were combined in the late 18th century in England to yield large water- or steam-driven spinning machines. The ring-spinning machines shown in the photograph were the commonest type of spinning frame in the U.S. in the early 20th century. All the machines in the spinning room were driven by a single electric motor; power was transmitted to the machines by the belts extending from the ceiling. The upright bobbins at the top of the machine held the crude ropelike material called roving. The strand of roving was drawn down between two leather-covered cylinders and through a loop of wire called the traveler. The traveler moved on a ring circling the yarn bobbin. The yarn bobbins are at waist level; each machine had about 300. The traveler revolved about the bobbin some 10,000 times per minute, stretching and twisting the yarn as the ring moved up and down distributing the yarn evenly. In the early mills most of the women who operated spinning and weaving machines came from farm or artisan families. Such women had customarily worked before marriage; the opening of the mills merely changed their place of employment.

in the 1830's: "At first the sight of so many bands, and wheels, and springs in constant motion was very frightful. She felt afraid to touch the loom, and she was almost sure that she could never learn to weave.... The shuttle flew out and made a new bump upon her head; and the first time she tried to spring the lathe, she broke out a quarter of the threads."

Industries other than textiles also began to draw workers from the pool of young single women, including the manufacturers of paper, buttons, shoes and watches. Later in the century the makers of electric wire and light bulbs recruited labor from the same source. Images of the factories where such goods were made capture the novelty of the workers' experience. In engravings and photographs rows of young women stand at attention before their machines. The depiction of many employees with identical posture, dresses and hair styles conveys the scale of the enterprise. The images give female labor a uniform and impersonal quality that was then strange. The comparison implied in such images is with the more intimate surroundings of the household.

Those who argue that the new techniques of manufacturing had a revolutionary impact on women assume that factory work drew women permanently from their traditional place in the home. Implicit in the argument is the notion that women did not work for wages or engage in other productive activities before the Industrial Revolution. The opportunity to earn wages, it is thought, gave women entry into the world of men, where they found independence and social recognition. In reality women's work in the early factories was conceived by employers (and to some extent by the women themselves) in traditional terms. The most important of these was that wage work was a secondary occu-

TELEPHONE EXCHANGE was a place of employment for increasing numbers of young women in the early part of the 20th century, as is indicated by this photograph of the central exchange of Kansas City, Mo., in 1904. The switchboard shown is of the type called a multiple switchboard. Introduced in 1897, it incorporated numerous advances over its predecessors. The first switchboards were cumbersome: the signaling operations (such as the customer's indicating to the operator that he wanted to make a call) were handled by equipment separate from that utilized for calling. All the circuits were incorporated in a single board. Much of the subsequent improvement resulted from the invention of the switchboard jack, a small socket that carried current for both signaling and calling. The call was put through by connecting two jacks with a short cord that had a plug at each end. A version of the jack was patented in 1879. By 1897 the jack had been made small enough so that 10,000 lines could be put within the reach of one operator. A further advance was to separate the jacks of callers ("answering jacks") from those of the people to be called ("connecting jacks"). Each operator had in front of her on three

pation and that a woman's real work was raising children and running the household. The traditional conception was demonstrated in the employers' preference for hiring women who were young and unmarried: most machine operatives were between the ages of 16 and 25.

That factory work was an extension of previous experience is further demonstrated by the fact that the operatives came largely from artisan and agricultural families; such families had for generations expected their daughters to work at home, in a cottage industry or in domestic service until they married. In both Europe and the U.S. there was substantial variation among regions and social groups in the fraction of young single women who earned wages. In spite of the variation the fact remains that in Europe and the U.S. many young women worked outside the home. When the mills opened, they offered better pay and more jobs than had previously been available to young women, who merely shifted their employment to a new place.

Furthermore, a job in the mill did not alter the anticipated course of a woman's life by substituting paid employment for marriage and the care of children. Mill work simply provided young women with a new kind of job at a time of life when they would ordinarily expect to be employed. Most operatives took jobs for immediate financial gain and not in the hope of a career. Once a woman was married, unless her family desperately needed her wages there was little reason for her to spend her adult life as a mill worker. The possibility of promotion and higher pay was remote. The majority of women therefore left the mill when they married or when they later had children.

The work force in the mill therefore

vertical panels connecting jacks for every subscriber in the system. On a horizontal shelf she had answering jacks for only a fraction of the subscribers. To place a call the customer lifted the receiver, activating a circuit that lighted a lamp above the answering jack. The operator inserted one plug in the answering jack; the lamp went out and power for voice transmission was supplied from a battery in the telephone-company building. The operator pressed a key, thereby connecting her headset to the circuit, and asked for the number to be called. If the desired line was free, she put the other plug into the connecting jack and pressed a key that rang the telephone bell of the person being called. Lights on the switchboard indicated when the call was answered and when either person hung up. The first commercial telephone exchange was established in New Haven, Conn., in 1878 with 21 subscribers. In the earliest exchanges the operators were young men, but women replaced them in the 1880's. Work as a telephone operator was clean and respectable and therefore thought to be suitable for the young single middle-class women who were beginning to work in substantial numbers in the late 19th century.

consisted largely of single women, with a minority of poor married women and widows. The turnover in the work force was high, but employers did not object because the turnover did not hamper production. Women learned their tasks quickly, and they accepted the conditions in the factory, in part because they did not expect to stay there. From the point of view of the mill owner whatever drawbacks there may have been in a constantly changing female labor force were outweighed by one great benefit: such a labor force was cheap.

One keen observer of the process of industrialization in England was a Scottish professor, Andrew Ure. Like many of his contemporaries he was fascinated by the emerging industrial processes; he traveled through England recording his observations of factories. In 1835 he noted: "It is in fact the constant aim and tendency of every improvement in machinery to supersede human labour altogether or to diminish its cost, by substituting the industry of women and children for that of men.... In most of the water-twist or throstle cotton-mills [a throstle was a large spinning frame driven by water power], the spinning is entirely managed by females of 16 years and upwards. The effect of substituting the self-acting mule for the common mule is to discharge the greater part of men spinners and to retain adolescents and children. The proprietor of a factory near Stockport states that by such substitution, he would save £50 a week in wages, in consequence of dispensing with nearly forty male spinners at about 25*s* of wages each."

Ure's observation reflects the longstanding belief that women did not merit or require wages as high as those of men. Because of exclusion from trades practiced by men, a lack of training and the assumption that women's wages would supplement the family income rather than provide it, the prevailing evaluation of women's work was that it was worth less than that of men.

Not only were women in the mill paid less than men for the same work; cultural attitudes about women's capacities also led to the designation of many jobs as being suitable primarily for women. Employers hired women as mill operatives, they said, because their small, graceful fingers could piece the threads together easily. In addition the female temperament—passive, patient and careful—was thought to be perfectly suited to boring, repetitive work. Men were employed in the mills as supervisors, mechanics and occasionally as operatives in such tasks as carding, which required considerable strength. Women tended spinning, winding, warping and weaving machines. The specific jobs done by men and women varied from mill to mill, but the separation of male and female work was almost universal; in most mills many rooms were staffed entirely by women. Thus the pattern of

CLERICAL WORK was transformed between 1880 and 1910, in part by the introduction of the typewriter. This photograph depicts part of the Audit and Policy Division of the Metropolitan Life Insurance Company in about 1910. It reveals one of the results of the transformation: women had replaced men as clerical workers. In the Audit and Policy Division the reports of field agents were audited and insurance policies were written. In 1910 the division had more than 500 women clerks who worked at the typewriter, which had become a practical instrument of office work. The first workable typewriter was made in 1867 by Christopher Latham Sholes. Sholes's machine was put on the market in 1873 by E. Remington & Sons, the gunsmiths, as the Remington No. 1. It had many of the features of the modern manual typewriter, including a cylinder with line-spacing and carriage-return mechanisms, an escapement for the spacing of letters, and type bars that struck at a common point on the cylinder. The keys were in an arrangement much like that of the modern machine. Two further advances were required to yield the typewriter of 1910. One was the shift key, which made it possible to type both capital and lowercase letters on a single keyboard. (The Remington No. 1 could type only capital letters; some contemporaneous designs had a second keyboard for the lowercase letters.) The other advance was the placement of the paper so that the work could be read while it was being done. (In most of the early machines the type bars had struck at a point on the underside of the cylinder and the carriage had to be lifted for the work to be read.) In 1910 women were relatively new participants in clerical work. Earlier, young men had done such work in preparation for administrative positions. The growing volume of paperwork in the 1880's and 1890's resulted in the creation of large numbers of secretarial jobs with little opportunity for advancement. Because the new jobs did not offer the possibility of a career they were thought by employers to be suitable for women. The segregation of women in such work is indicated by the fact that in 1908 the Audit and Policy Division had 287 bookkeepers, all of whom were men, and 752 clerks, all of whom were women.

separate realms of work for men and women remained undisturbed. (The notion of separate spheres of work for men and women is so deeply entrenched in cultural images that the division is presumed to extend to the first workplace, where "Adam delved and Eve span.")

Wages for women lower than those of men and the segregation of jobs according to sex were often the result of mechanization in the 19th century and the early 20th. Machinery that extended the division of labor, simplified and routinized tasks and called for unskilled workers rather than skilled craftsmen was usually associated with the employment of women. From the point of view of the skilled workers displaced by machinery feminization meant the devaluation of their work.

The increase in female white-collar office work at the end of the 19th century was a new variation on the theme. The telephone and the typewriter have come to symbolize the reorganization of clerical work at that time. These innovations, however, were only a small part of the reorganization. Increases in the urban population and manufacturing and the consequent expansion of commerce called for enormous amounts of paperwork. Earlier in the 19th century young men had done clerical work as part of a general apprenticeship in business; such apprenticeships were often preparation for partnership or inheritance of the enterprise. As the volume of paperwork increased, however, clerical work was separated from administrative work and from advancement in the executive hierarchy.

In the early phase of the development of the modern office, copy work was given out to women to be done at home. In the U.S. such workers were usually married women or widows with children who supplemented the household income by copying; they were paid by the word. The literacy of the copy workers indicates that they were educated women and therefore probably from artisan or even middle-class families.

The first phase of modern office work did not last long. The first practical typewriter, invented in 1867, was introduced into commercial use in the 1870's and quickly became standard office equipment. Typing, stenography and filing became components of a full-time job done in the office. As the role of secretary was created ambitious young men moved into sales, advertising and administrative positions. Women were hired in the new white-collar service jobs. The shift took only a few decades. In the U.S. census of 1880 only a few women were listed as office clerical workers. By 1910, 83 percent of all stenographers and typists were women; the proportion was similar in France and England. The feminization of clerical work has continued: in 1980, 97 percent of typists in the U.S. were women, as were 89 percent of stenographers.

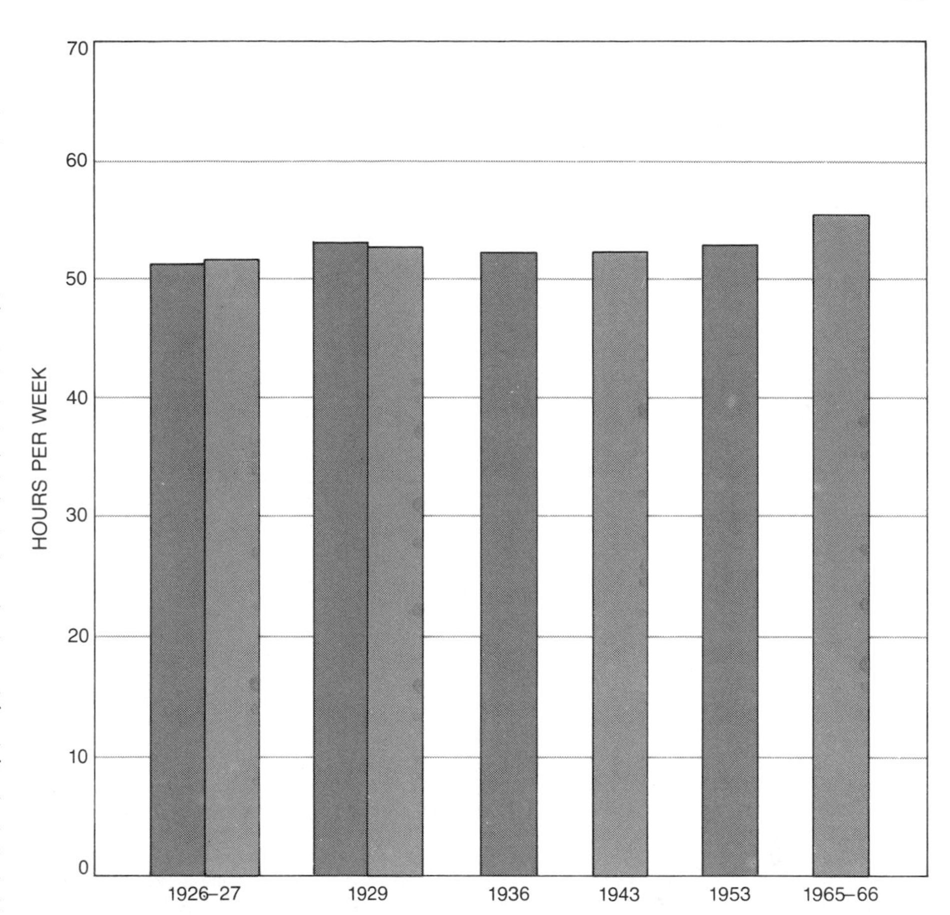

TIME SPENT IN HOUSEWORK each week by women not employed outside the home changed little between 1926 and 1966. Gray bars are for rural women, colored bars are for urban women. The data are from various comparable surveys. The period was one in which the urban population grew and the ownership of household appliances became widespread. It has been argued that such appliances freed women to do other kinds of work. By the late 1960's, however, women not employed outside the home still spent more than 50 hours per week in housework. Women who were employed also spent a substantial amount of time in housework: 26 hours per week. It appears that rather than being freed for other kinds of work by household appliances many women went to work largely in order to be able to buy such appliances.

Hence like the spinning and weaving rooms of the textile mill the outer office quickly became a feminine space. Office work, however, had qualities that distinguished it from blue-collar work. It called for some formal education. Furthermore, because it was clean, respectable work it was thought to be suitable for middle-class women who had not previously worked for wages. Families who wanted relief from supporting a single daughter or who sought to give marketable skills to daughters who might not marry or whose husbands might die young sent their daughters to commercial schools and then into the job market. In the 1870's such economic and demographic pressures had propelled young middle-class women into nursing and teaching. In the 1890's and 1900's the pressures moved them toward the newly available office jobs. In the offices they were joined by women from poor families who had gone to commercial training schools to acquire the skills for a white-collar job.

The secretary (once widely known as the "female typewriter") and the telephone operator quickly replaced the machine operative as the typical female worker. Their work was much less dirty and less difficult than mill work. There were, however, fundamental similarities between the situation of the blue-collar workers and the white-collar ones. The mechanization of document copying and communications created new occupations while maintaining women in a labor market separate from that of men. Occupations were still segregated according to sex, and the cultural stereotypes of women's capacities were closely associated with the work. It was said that women's fingers raced as deftly over the typewriter keys as if they had been playing the piano. According to employers, women's ability to greet strangers pleasantly, their reliability and their tolerance for repetition made them ideal telephone operators.

Just as jobs in the mill had been, so the jobs of secretary and telephone operator were designated as employment for single women. Age limits of between 18 and 25 were usually enforced. Em-

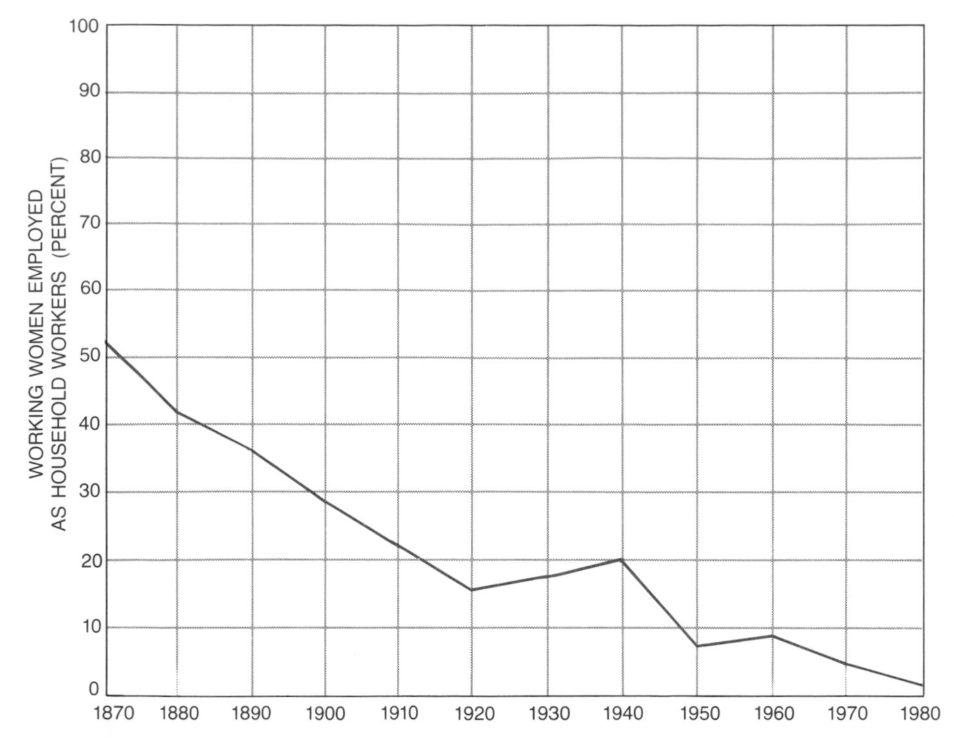

HOUSEHOLD WORKERS (other than family members) represent a diminishing fraction of all working women. In 1870 more than half of all employed women were household workers; by 1980 the fraction had decreased to about 2 percent. The reduction in the availability of domestic servants, along with higher standards of household cleanliness and the presence of laborsaving appliances, has served to increase housework for middle-class women. With the aid of appliances such women now spend much time doing work that was once done by servants.

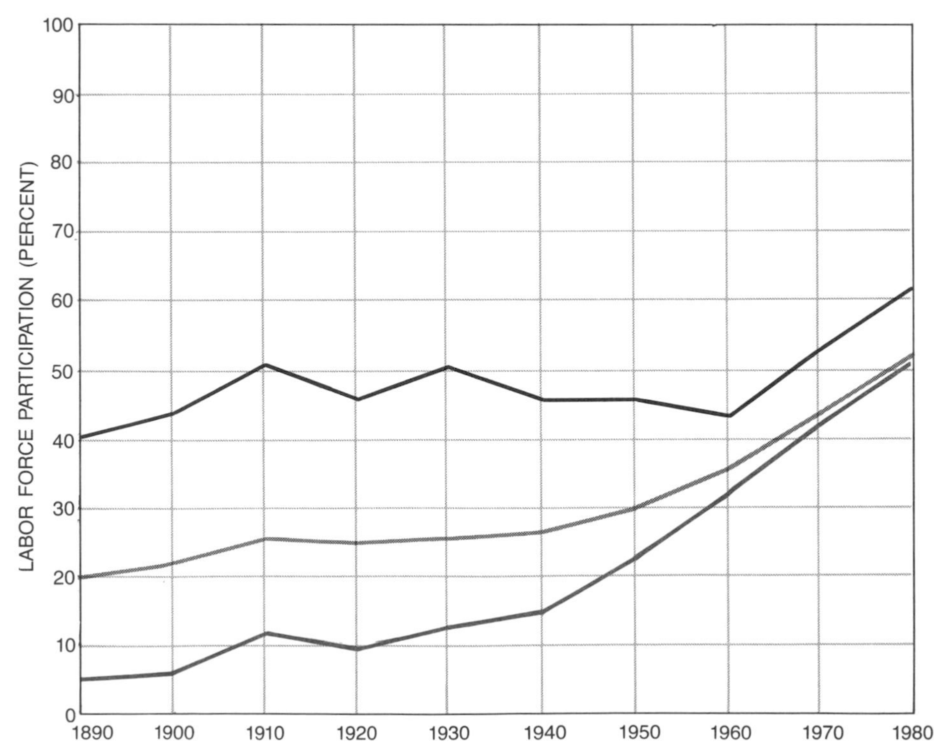

FRACTION OF MARRIED WOMEN WHO WORK has increased greatly since 1900. The black line shows the labor-force participation rate for single women, the gray line the rate for all women, the colored line the rate for married women. About half of all married women now work outside the home, compared with about 5 percent in 1900. In the 19th century and the early 20th most working women were single. Work did not change a woman's anticipated life course, because women stopped working when they married. The increase in the employment of married women was a substantial social change, but it appears to have had little directly to do with mechanization. One interpretation is that married women became acceptable employees only when the pool of single women did not expand rapidly enough to meet labor needs.

ployers often required young women to leave their job when they married whether they wanted to leave or not. A businesswoman who ran a commercial training school for female office workers early in the 20th century explained the difference between the careers and wages of men and those of women as follows: "Women must admit to one handicap in an independent business life—business wears a temporary aspect to most girls. For if she is normally constituted, every girl hopes that someday she will be happily married."

The temporary nature of employment made promotion and an institutionalized career path unnecessary; because women in business were not promoted regularly their wages remained low and stable. In addition employers assumed, although it was not always true, that young women did not have families to support and indeed that they were supported by their families. As a result the wages of female clerical workers were generally about half what male clerical workers had earned. For this reason men denounced the invasion of the office by women. A male former clerk wrote that "women are employed not on account of their capacity but because they are cheaper than men."

In spite of a substantial change in the composition of the female labor force working women are still paid less than their male counterparts. Since World War II there has been a dramatic increase in the proportion of married women who work; the increase has been particularly marked among those with young children. The increase has had little to do with mechanization, including, as I shall show, the mechanization of the household. No major technological innovation appears to have been directly associated with the increase in the number of married women who earn wages. Rather than being the result of mechanization the increase has been caused by a series of economic and demographic developments that have drawn married women into the kinds of jobs once held by single women.

Valerie Kincaid Oppenheimer of the University of California at Los Angeles has argued that married women became acceptable employees when the pool of single women workers decreased as a result of extended education and higher marriage rates. Over the same period inflation and the desire to maintain an increasingly high standard of living led many married women to seek work outside the home. Their motives, like those of the married women who worked in the early textile mills, were economic. For many 20th-century women, however, the immediate aim was not to secure food for the family but to pay off a mortgage, send children to college or buy laborsaving appliances. In the late

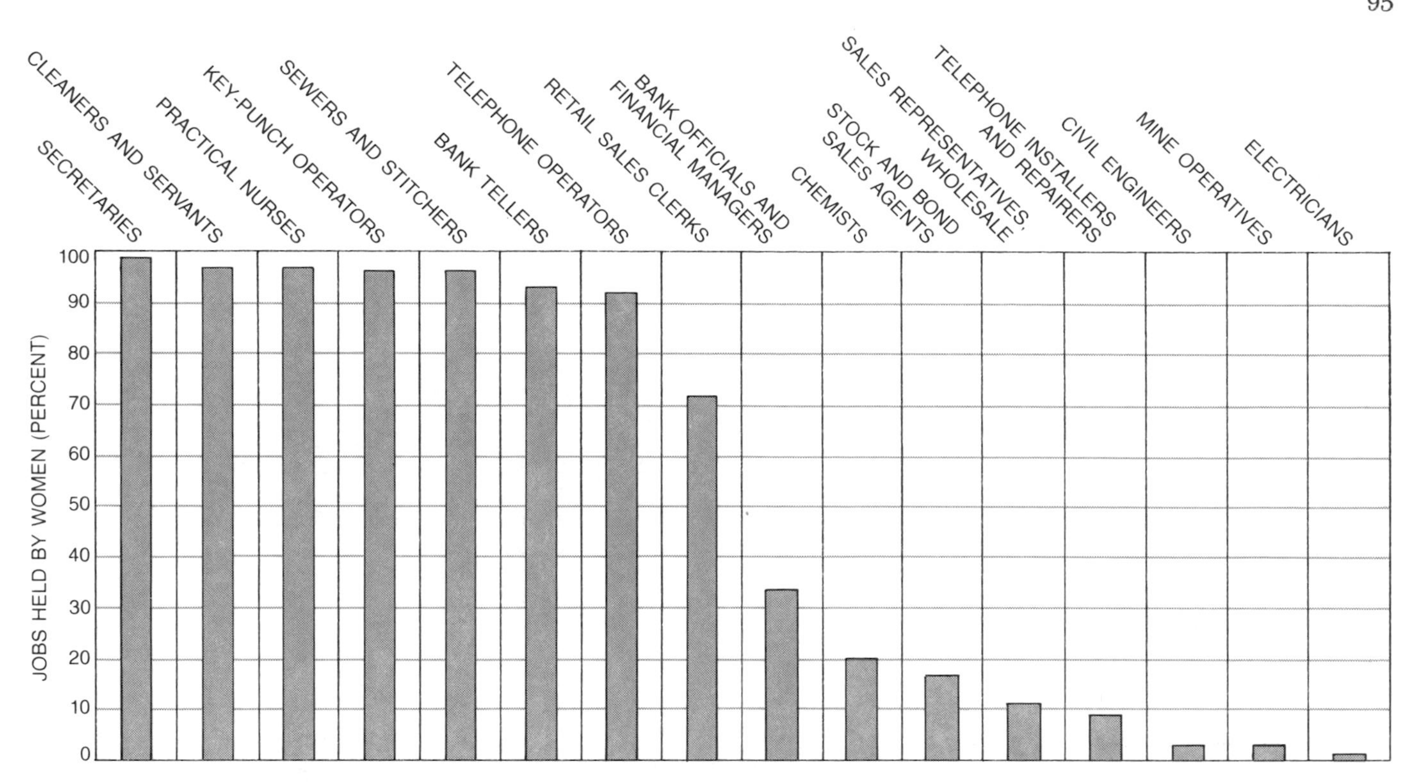

SEGREGATION OF OCCUPATIONS BY SEX has not been eliminated by the mechanization of work. This chart shows the fraction of workers in selected occupations who are women. The data are for 1980; they are from the U.S. Department of Labor. The occupations dominated by women are for the most part unskilled and poorly paid. Those dominated by men are a mixture of well-paid skilled jobs and blue-collar jobs. There are sharp distinctions between women's work and men's work even within a particular industry or between closely related occupations. In such cases men dominate the better-paid occupations. For example, the occupation of telephone operator, which became women's work in the 1880's as a result of mechanization and other factors, has remained women's work: in 1980 more than 90 percent of operators were women. Telephone installers and repairers, however, were overwhelmingly male. In 1980 more than 70 percent of retail sales clerks were women. Most wholesale salesmen, however, were men; only about 10 percent were women.

1970's and early 1980's, as the inflation rate and the divorce rate have increased, the older subsistence motive has reappeared. Many mothers now work in order to feed and clothe their children rather than to buy luxuries. The extraordinary number of families headed by women that live at or below the poverty level in the U.S. demonstrates that this is so; it also reflects the persistence of the attitude that women's work deserves less pay than men's work.

Although increased employment outside the home by married women was a significant shift, some important qualities of women's work remained unaltered when married women went to work. Manufacturing and white-collar jobs are still segregated by sex. In both kinds of employment men's and women's working areas are often separated spatially. Thus in modern workplaces there are not only men's and women's jobs but also men's and women's spaces, in the same sense in which the early telephone exchange was a women's space.

Observation of the effect on marriages and on children of women's leaving home to work has led some economists and employers to argue that if wage work is necessary, it ought to be done in a married woman's traditional space: the home. In the case of office work, which now employs the largest proportion of working women, electronic equipment could make it possible for married women to work at home [see "The Mechanization of Office Work," by Vincent E. Giuliano, page 75]. Connected by telephone lines to administrative headquarters, workers could operate word processors, retrieve and store information and do various other clerical and secretarial tasks. Mechanization might thereby reconcile work and child care for many women. It could also make it unnecessary for social planners to consider whether the nuclear family is the best form of organization of biological reproduction and child care.

It seems unlikely, however, that computer technology will transform the enduring characteristics of women's work. On the contrary, computer terminals in the home would probably lower secretarial wages. The reason is that the machines themselves make it possible to do large amounts of work quickly and therefore reduce the number of workers needed to do a given amount of work. The resulting increase in competition for jobs would drive wages down. Moreover, it is probable that when home terminals become common, employers will substitute piece rates for hourly wages or salaries. Piece rates offer a more efficient means of controlling unsupervised work than payment according to time worked. The isolation of office workers in the home would make it difficult for them to discuss shared grievances, as workers now can in the workplace, and hence would make it difficult for them to organize collectively. As a result it would be possible for employers to pay low and unequal wages.

By enabling married women to work at home for wages computers might have an effect on women analogous to that of the sewing machine in the 19th century. Sewing machines made needlework much more efficient. Initially, however, the machines were installed in the workshops of clothing manufacturers or subcontractors. In the 1890's the manufacture of lighter and cheaper models made sewing machines practical for home use. Along with the purchase of a sewing machine often went wage work at home. Advertisements for sewing machines sometimes included a contract with a garment manufacturer as an inducement to buy. If a woman signed a contract, she could by doing piecework both pay for her sewing machine and earn additional money.

Work at home was not a new occupation for married women. Women of the urban working class in Europe and the U.S. had long helped to support their family by means of such work. In some instances full-time entrepreneurial work overlapped the home, as in the case of women who were innkeepers;

other women did metal polishing and laundry in the home; still others did piecework such as making hats or artificial flowers or spinning silk.

The commonest occupation for married women at home, however, was sewing. The practice of earning wages by sewing at home became more widespread with the growth of the ready-made clothing industry. In the best circumstances married women combined sewing with child care and housework. Their earnings supplemented those of a husband and in some cases working children. The fact that the money earned in this way was supplemental to the family income enabled the women to exercise some control over the rhythm of their work. Sewing for a few hours a day was profitable employment while the children were at school.

Such relatively fortunate women were in the minority; most women sewed because they needed as much money as they could possibly earn. Because employers paid by the piece and the rates were low, long hours were needed to earn even a subsistence wage. Some working-class women spent every waking moment sewing and required the labor of as many family members as were available. It was common for children to be kept home from school so that more garments could be sewn. The sewing machine transformed such households into miniature sweatshops. The mother ran the machine while children and relatives sewed hems and put on buttons. Neighbor women who could not afford their own machines sometimes joined the work force, bringing young children to sleep or play while they worked. Social reformers at the turn of the century described tenement rooms filled with women and children whose voices could barely be heard over the noise of the sewing machine. The reformers deplored the low piece rates that led women to work 15 or more hours per day, neglecting their children and household.

The sewing machine increased the speed with which goods could be sewn, standardized the products and perhaps created more jobs. The machine did not, however, alter the low rate of pay, the fact that most home workers were women and the fact that most married working women worked at home. Thus the mechanization of needlework did not free working-class women from the household; instead the sewing machine was incorporated into the traditional pattern of work at home.

The sewing machine had a different effect on women who did not use it as a means of earning money. Housewives who had previously bought clothing ready-made began to make clothes at home with the aid of the patterns that were printed on the women's pages of newspapers and magazines and sold in fabric stores. The practice of making the family's clothes had diminished in importance with the introduction of mass-produced garments. By encouraging women to return to the older practice the mechanization of sewing reduced the time a housewife spent as a consumer and increased the time she spent as a household worker.

The sewing machine was one of several devices that "industrialized" the middle-class household in the first decades of this century. The washing machine, the iron and the home freezer also reduced the urban housewife's reliance on services provided outside the home. The new appliances individualized the preparation and preservation of food and the making and maintenance of clothing and linen. Instead of buying services, the housewife did the job herself with the aid of her appliances. The vacuum cleaner and the dishwasher had a similar effect, but those appliances did work that had been done by servants in most middle-class homes.

Some observers of mechanization have suggested that there is a causal relation between the industrialization of the household and the entry of married women into the labor market. It is usually argued that household appliances so diminished the time housewives needed for domestic chores that they had time for paid work outside the home. John D.

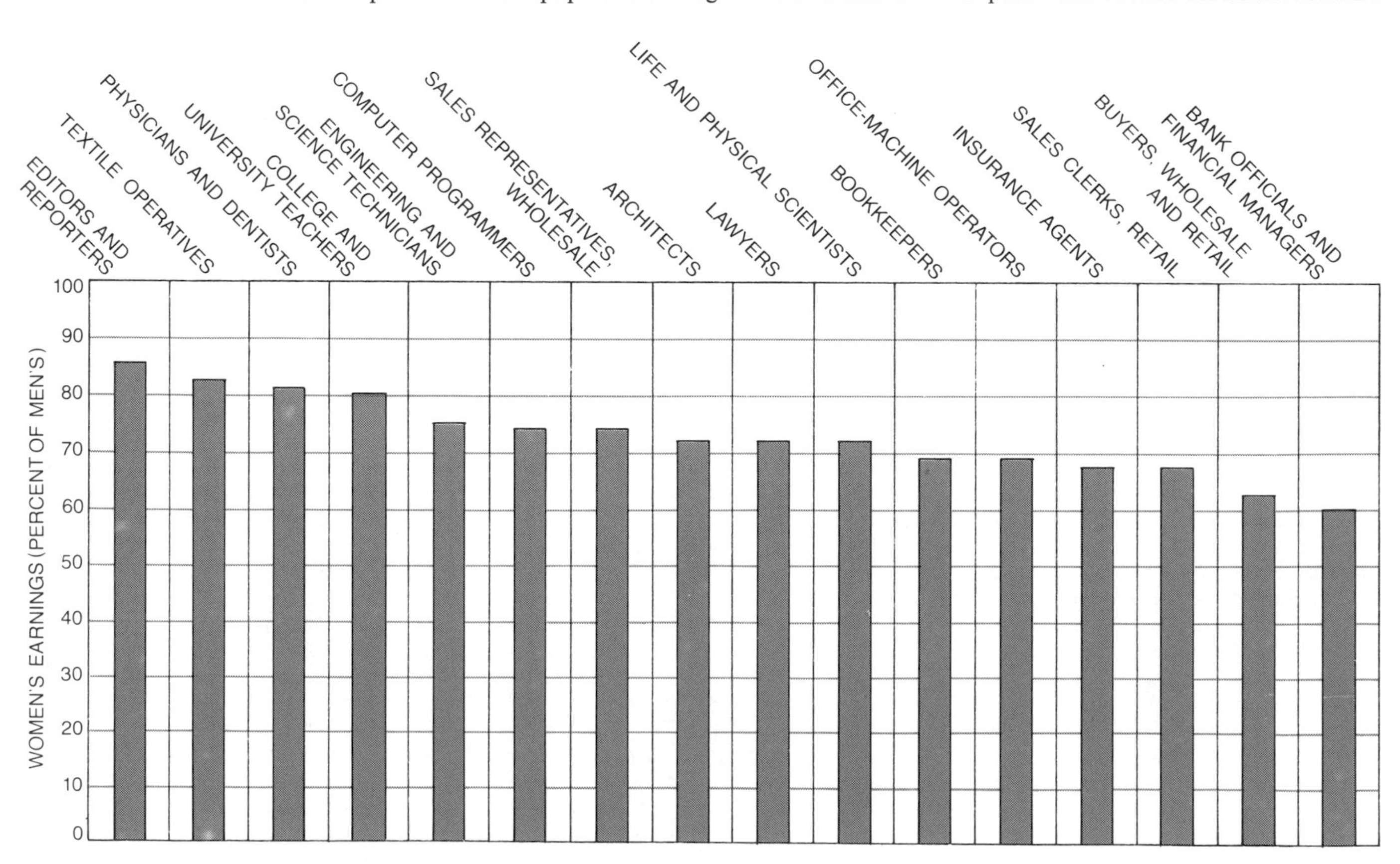

WOMEN'S EARNINGS are less than those of men in the same occupation in almost every field. The bars indicate women's earnings as a fraction of men's. The data are for 1981; they are from the U.S. Department of Labor. The differential between the earnings of men and the earnings of women varies from about 40 percent among bank officials to about 15 percent among editors and reporters. The differential exists in both skilled and unskilled occupations; it exists in occupations dominated by men and those dominated by women.

Durand of the Bureau of Social Affairs of the United Nations predicted in 1946 that laborsaving household appliances might "virtually ... eliminate the home as a place of work and housewives as a functional group in the population." The work of Joann Vanek of the UN Statistical Office, however, shows that between 1920 and 1960 the time women not employed outside the home spent in housework increased [see "Time Spent in Housework," by Joann Vanek; SCIENTIFIC AMERICAN, November, 1974]. The four decades covered by Vanek's study constitute a period when the urban population of the U.S. increased and the ownership of household appliances spread throughout the population. In spite of the fact that rural women preserved food and did much more physical work than urban women, Vanek's investigation shows that urban housewives spent more time in housework than rural ones.

One reason for the increase in the time spent in housework appears to have been the decrease in the proportion of urban families that had domestic servants. By the 1920's the population of servants had declined both in number and as a proportion of the female work force. In 1870, 52 percent of employed women were servants; by 1920 the fraction had decreased to 16 percent. Middle-class women without servants found vacuum cleaners and washing machines an attractive investment, but although the appliances eliminated some of the more onerous aspects of housework, they did require someone to operate them. In the household without servants the housewife became the sole domestic worker.

The amount of domestic work was further increased by a rise in the standards of household cleanliness. Ruth Schwartz Cowan of the State University of New York at Stony Brook has shown that this rise accompanied the introduction of household appliances. From the 1890's to the 1920's the home-economics movement portrayed women as scientific managers of the health of the household. Articles in magazines such as *Ladies' Home Journal* emphasized the importance of spotless homes; advertisements in their pages offered soaps and cleaning solutions to help do the job. Even if women spent less time sweeping, scrubbing and rinsing than their mothers and grandmothers had, they did the laundry more often and spent more time waiting in line at stores to buy cleaning agents.

The time spent in domestic chores was also increased by a new emphasis on the principles of child rearing. Mothers were expected to be experts in the psychological, physical and educational development of their children. Household appliances merely enabled women to transfer their attention from one kind of domestic activity to another without by any means escaping such activity altogether. Cowan concludes that "women of the middle class ... did not get divorces, nor enter the labor market or the political arena. They were too busy sterilizing bottles, taking children to music and dance classes, making balanced meals, shopping, studying child psychology and sewing color-coordinated curtains."

CONTEMPORARY TELEPHONE OPERATORS are shown in a photograph made in a Traffic Service Position System (TSPS) facility operated by the New York Telephone Company in New York. Introduced in 1969, the TSPS consists of a base unit where most calls are automatically connected and offices such as this one, where operators handle calls that require human intervention. The TSPS made it possible for customers to dial calls requiring the assistance of an operator, such as credit-card calls. The customer dials the call with the addition of a code indicating that assistance is required. The operator then gets the customer's credit-card number and enters it in the record of the call. The automatic completion of the calling circuit, which eliminated the need for an operator on routine calls, was introduced relatively slowly; by 1925 only about 12 percent of the telephone lines in the Bell System were operated automatically. Thereafter the expansion and mechanization of telephone service were rapid; thus the number of operators increased slowly. In the 1910's the Bell System had 100,000 operators for seven million telephones. In 1970 there were 166,000 operators for 98 million telephones.

The hypothesis that laborsaving machinery liberated women from domestic work is further weakened by the observation that although middle-class women were the most likely to be able to afford household appliances, they have always been the group least likely to work for pay outside the home. There is little historical correlation between the owning of household machinery and wage earning. In the early part of the century increases in the female labor force were due primarily to single women taking white-collar positions; the increase was due to a lesser extent to poor married women filling less desirable jobs in factories and in domestic service.

The fact that the women who were working were not the ones who owned the new appliances was demonstrated in the stark contrast between the homes of the middle class and those of the working class. Investigators such as Jacob Riis of conditions in the homes of the poor found them lacking in essential comforts and indeed barely furnished. In 1912 an investigator for a Senate committee wrote that "nothing appears comfortable, nothing beautiful." Working-class homes were kept clean by old-fashioned labor-intensive methods. The women of such households were not freed to work by laborsaving appliances; they were forced to work by economic necessity.

Since World War II there has been a diffusion of household appliances to a large proportion of U.S. households, partly because the appliances cost less in relative terms than they had earlier. There has also been a dramatic increase in the number of married women who work, in both the middle class and the working class. The evidence nonetheless indicates that economic pressures and not free time propel most women into the labor force. If anything, it is the desire for household appliances rather than their possession that provides some of the motivation for women to work. Their goal is to earn enough money to buy appliances that promise to lighten the double burden of wage work and housework. The evidence confirms Oppenheimer's suggestion that "the great increase in laborsaving devices and services [is] a response to a rise in female labor participation" and not the cause of wage-earning work.

Although the ownership of household machinery can lighten a working woman's domestic responsibilities, it does not by any means eliminate household work. Married working women continue to reconcile domestic work and wage earning as their predecessors did in the preindustrial era. Data collected in the 1970's show that working women who are married spend an average of about 30 hours a week on housework compared with 50 for full-time housewives. Thus although the burden of housework is reduced somewhat, working women still spend a significant amount of time on domestic and child-rearing tasks. In contrast, the husbands of working wives spend little or no time doing housework. Many married women appear to be trying to manage the combination of wage earning and housework by taking part-time rather than full-time jobs, which perpetuates occupational segregation and the lower status of women in the job market.

In spite of the fact that the past 200 years have constituted a period of rapid technological change there is a surprising continuity between the social and economic position of women at the beginning of the period and at the end. In both blue- and white-collar work mechanization has been associated with the feminization of particular occupations. Industrial machinery has been introduced partly to lower labor costs; employers have drawn on long-standing cultural assumptions about the low value of women's work in designating certain jobs as suitable only for females. The mechanization of the household confirmed the housewife's responsibility for preparing food, buying supplies and keeping things clean in a private family setting. Indeed, mechanization has made these responsibilities more acceptable to middle-class women who would once have relied on servants. There have been changes in the location of women's work and in the level of drudgery involved, but the changes have not been revolutionary. There is a predictable pattern in the changes in women's work: each transformation has extended the notion of a location for women's work separate from that for men's work and the notion that women's work is worth less than men's.

My argument is not intended to deny that there have been significant improvements in certain aspects of the position of women in society since the mechanization of work began. In the U.S. and western Europe laws enacted in the 19th century granted women education and property rights; laws enacted in the 20th century granted them the right to vote. Social custom has altered standards of propriety in clothing, public behavior and sexual expression. Women now appear to have more choices and to be freer of repression and control than they were even a generation ago. There are more women training to become lawyers, physicians, university professors or business executives.

Whether such changes alter fundamentally the structure of society or women's position in it is debatable; in any case they are not the direct result of mechanization. Some changes may be indirect consequences of the Industrial Revolution. Industrialization accelerated the decline in the importance of landed property as the basis for family power. It led to a growing need for educated workers of both sexes and hence for teachers, and it stimulated the growth of the urban population, which needs myriad commercial services and health care. Other changes in the status of women have resulted from the introduction of more reliable methods of contraception; still others result from shifts in the ages at which people marry, bear children and die.

The most important improvements in the position of women, however, have been the result of the actions of women themselves. Such changes have sometimes been responses to mechanization, but they were not inherent in the machines. For example, textile factories brought women workers together under one roof and gave them a sense of the collective power of labor. In so doing the mills created the preconditions for organized expression of labor grievances. Demands for improved conditions and higher wages, however, were formulated by the women themselves; the demands were won only because of the economic and political pressure the organized workers brought to bear on their employers.

Education and, among the middle class, employment led women to demand civil equality, particularly the right to vote. Women won their civil rights, however, only after years of political organizing and (in England and the U.S.) militant public action. Household appliances may have made domestic workers out of middle-class women and so led them into the women's movement in the 1960's and 1970's, but it was the women themselves who formulated the critique of "the feminine mystique." These examples suggest that mechanization did not change the inferior position of women. On the contrary, mechanization emphasized women's social inferiority and led to protests aimed not only at improving particular conditions but also at improving the overall situation of women.

Those who insist that only a revaluation of women's status can lead to greater economic equity and the integration of women into all sectors of the labor market address the problem directly. Until the social and cultural conception of the value of women's work has been changed there can be no revolutionary transformation of women's status as workers. The mechanization of work affects those who work and society at large only through the social context in which the machinery is employed. For women mechanization has confirmed rather than altered their economic and social valuation. In spite of the political and industrial revolutions of recent centuries the revolution for women is yet to come.

8

THE DISTRIBUTION OF WORK AND INCOME

The Distribution of Work and Income

by WASSILY W. LEONTIEF

When workers are displaced by machines, the economy can suffer from the loss of their purchasing power. Historically the problem has been eased by shortening the work week, a trend currently at a standstill

"My Lords: During the short time I recently passed in Nottinghamshire not twelve hours elapsed without some fresh act of violence;... I was informed that forty Frames had been broken the preceding evening. These machines... superseded the necessity of employing a number of workmen, who were left in consequence to starve. By the adoption of one species of Frame in particular, one man performed the work of many, and the superfluous labourers were thrown out of employment.... The rejected workmen in the blindness of their ignorance, instead of rejoicing at these improvements in art so beneficial to mankind, conceived themselves to be sacrificed to improvements in mechanism."

With these words Lord Byron in his maiden speech to the House of Lords in February, 1812, sought to explain, and by explaining to excuse, the renewal of the Luddite protest that was shaking the English social order. Nearly a generation earlier Ned Ludd had led his fellow workers in destroying the "frames": the knitting machines that employers had begun to install in the workshops of the country's growing textile industry. The House had before it legislation to exact the death penalty for such acts of sabotage. The Earl of Lauderdale sharpened Byron's thesis that the misled workers were acting against their own interests: "Nothing could be more certain than the fact that every improvement in machinery contributed to the improvement in the condition of persons manufacturing the machines, there being in a very short time after such improvements were introduced a greater demand for labour than ever before."

History has apparently sustained the optimistic outlook of the early exponents of modern industrial society. The specter of involuntary technological unemployment seems to remain no more than a specter. Beginning with the invention of the steam engine, successive waves of technological innovation have brought in the now industrial, or "developed," countries a spectacular growth of both employment and real wages, a combination that spells prosperity and social peace. Thanks as well to technological innovation, more than half of the labor force in all these countries—70 percent of the U.S. labor force—has been relieved from labor in agriculture and other goods-production that employed substantially everyone before the Industrial Revolution. It is true that the less developed countries are still waiting in line. If the outlook for the future can be based on the experience of the past 200 years, those countries too can expect to move up, provided their governments can succeed in reducing their high rate of population growth and desist from interfering with the budding of the spirit of free private enterprise.

There are signs today, however, that past experience cannot serve as a reliable guide for the future of technological change. With the advent of solid-state electronics, machines that have been displacing human muscle from the production of goods are being succeeded by machines that take over the functions of the human nervous system not only in production but in the service industries as well, as has been shown in the preceding articles in this *Scientific American* book. The relation between man and machine is being radically transformed.

The beneficence of that relationship is usually measured by the "productivity" of labor. This is the total output divided by the number of workers or, even better, by the number of man-hours required for its production. Thus 30 years ago it took several thousand switchboard operators to handle a million long-distance telephone calls; 10 years later it took several hundred operators, and now, with automatic switchboards linked automatically to other automatic switchboards, only a few dozen are needed. Plainly the productivity of labor—that is, the number of calls completed per operator—has been increasing by leaps and bounds. Simple arithmetic shows that it will reach its highest level when only one operator remains and will become incalculable on the day that operator is discharged.

STOCKHOLDERS' MEETING (*Hauptversammlung*) of Volkswagenwerk AG in West Germany exemplifies an institution of the

The inadequacy of this conventional measure is perhaps better illustrated if it is applied to assess the effects of the progressive replacement of horses by tractors in agriculture. Dividing the successive annual harvest figures first by the gradually increasing number of tractors and then by the reciprocally falling number of horses yields the paradoxical conclusion that throughout this time of transition the relative productivity of tractors tended to fall while the productivity of the horses they were replacing was rising. In fact, of course, the cost-effectiveness of horses diminished steadily compared with that of the increasingly efficient tractors.

In the place of such uncertain abstractions it is more productive to try to bring the underlying facts into consideration and analysis. Technological change can be visualized conveniently as change in the cooking recipes—the specific combinations of inputs—followed by different industries to produce their respective outputs. Progress in electromechanical technology enabled the telephone company to replace the old technological recipe calling for a large number of manual switchboards having many operators with a new recipe combining more expensive automatic switchboards having fewer operators. In agriculture technological progress brought the introduction of successive input combinations with smaller inputs of animal and human labor and larger and more diversified inputs of other kinds—not only mechanical equipment but also pesticides, herbicides, vaccines, antibiotics, hormones and hybrid seed.

New recipes come into service in every industry by a constant process of "costing out." Some inputs included in a new recipe are at the outset too expensive, and it takes some time before improvements in their design or in the method of their manufacture bring sufficient reduction in their price and consequently in the total cost of the recipe to allow the adoption of the new technology. The decline, at the nearly constant rate of 30 percent per year for many years, in the cost per memory bit on the integrated-circuit chip has brought solid-state electronics technology first into expensive capital equipment such as telephone switchboards, automatic pilots, machine tools and computers, then into radio and television sets and powerful, low-cost computers as an entirely new category of consumer goods, then into the control systems of automobiles and household appliances and even into such expendable goods as toys. Thus the adoption of a new recipe in one industry often depends on replacement of the old by a new technology in another industry, as the vacuum tube was replaced by the transistor and its descendants in the transformed electronics industry.

Stepping back and contemplating the flow of raw materials and intermediate products through the input-output structure of an industrial system and the corresponding price structure, one can see that prices more or less faithfully reflect the state of technology in the system. With the passage of time price changes can be expected to reflect long-run technological changes going on in the various sectors. In this perspective human labor of a specific kind appears as one, but only one, of the many different inputs the price of which must be reckoned in the costing out of a given

West German economy: close collaboration between capital and labor. The West German "codetermination" law requires that half of the board of directors of each large corporation be elected by labor and the other half by the stockholders. At this meeting of Volkswagen, held on July 1, the directors and the managers of the company are on the dais giving reports to stockholders and answering questions.

technological recipe. Its price, the wage rate, enters into the cost comparisons between competing technologies in the same way as the price of any other input.

In the succession of technological changes that have accompanied economic development and growth, new goods and services come on the stage and old ones, having played their role, step off. Such changes proceed at different rates and on different scales, affecting some sectors of economic activity more than others. Some types of labor are replaced faster than others. Less skilled workers in many instances, but not always, go first, more skilled workers later. Computers are now taking on the jobs of white-collar workers, performing first simple and then increasingly complex mental tasks.

Human labor from time immemorial played the role of principal factor of production. There are reasons to believe human labor will not retain this status in the future.

Over the past two centuries technological innovation has brought an exponential growth of total output in the industrial economies, accompanied by rising per capita consumption. At the same time, until the middle 1940's the easing of man's labor was enjoyed in the progressive shortening of the working day, working week and working year. Increased leisure (and for that matter cleaner air and purer water) is not counted in the official adding up of goods and services in the gross national product. It has nonetheless contributed greatly to the well-being of blue-collar workers and salaried employees. Without increase in leisure time the popularization of education and cultural advantages that has distinguished the industrial societies in the first 80 years of this century would not have been possible.

The reduction of the average work week in manufacturing from 67 hours in 1870 to somewhat less than 42 hours must also be recognized as the withdrawal of many millions of working hours from the labor market. Since the end of World War II, however, the work week has remained almost constant. Waves of technological innovation have continued to overtake each other as before. The real wage rate, discounted for inflation, has continued to go up. Yet the length of the normal work week today is practically the same as it was 35 years ago. In 1977 the work week in the U.S. manufacturing industries, adjusted for the growth in vacations and holidays, was still 41.8 hours.

Concurrently the U.S. economy has seen a chronic increase in unemployment from one oscillation of the business cycle to the next. The 2 percent accepted as the irreducible unemployment rate by proponents of full-employment legislation in 1945 became the 4 percent of New Frontier economic managers in the 1960's. The country's unemployment problem today exceeds 9 percent. How can this be explained?

Without technological change there could, of course, be no technological unemployment. Nor would there be such unemployment if the total population and the labor force, instead of growing, were to shrink. Workers might also hang on to their jobs if they would agree to accept lower wages. Those who are concerned with population growth are likely to proclaim that "too many workers" is the actual cause of unemployment. Libertarians of the "Keep your hands off the free market" school urge the remedy of wage cuts brought about by the systematic curtailment of the power of trade unions and the reduction of unemployment and welfare benefits. Advocates of full employment have been heard to propose that labor-intensive technologies be given preference over laborsaving ones. A more familiar medicine is prescribed by those who advocate stepped-up investment in accelerated economic growth.

Each of these diagnoses has its shortcomings, and the remedies they prescribe can be no more than palliative at best. A drastic general wage cut might temporarily arrest the adoption of laborsaving technology, even though dirt-cheap labor could not compete in many operations with very powerful or very sophisticated machines. The old trend would be bound to resume, however, unless special barriers were erected against laborsaving devices. Even the most principled libertarian must hesitate to have wage questions settled by cutthroat competition among workers under the pressure of steadily advancing technology. The erection of Luddite barriers to technological progress would, on the other hand, bring more menace to the health of the economic and social system than the disease it is intended to cure.

Increased investment can certainly provide jobs for people who would otherwise be unemployed. Given the rate of technological advance, the creation of one additional job that 20 years ago might have required an investment of $50,000 now demands $100,000 and in 20 years will demand $500,000, even with inflation discounted. A high rate of investment is, of course, indispensable to the expanding needs of a growing economy. It can make only a limited contribution to alleviating involuntary technological unemployment, however, because the greater the rate of capital investment, the higher the rate of introduction of new laborsaving technology. The latest copper smelter to go into service in the U.S. cost $450 million and

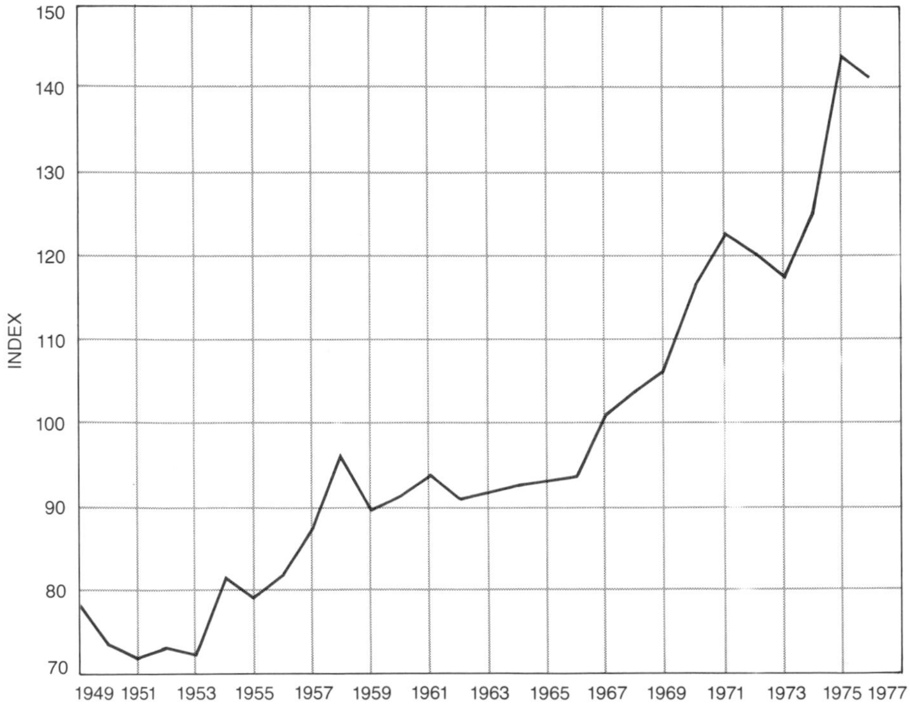

VALUE OF CAPITAL STOCK employed per man-hour in manufacturing industries in the U.S., plotted here on a constant 1967-dollar index, has almost doubled since the end of World War II. In this period the total output per capita of U.S. manufacturing industries also more than doubled. With a constant work week over this period (*see illustration on opposite page*) and an increase of only 4 percent in the blue-collar factory work force, from 12.8 to 13.3 million, the increase in output must be attributed almost entirely to the introduction of new technology embodied in the expanding capital stock of the industries. The development of the technology in these capital inputs is one of the functions of the white-collar "nonproduction" work force in the manufacturing industries, which in the same period more than doubled in number, from fewer than three million workers to nearly six million. The chart may also be taken as plotting the rising capital cost of creating a new job in U.S. manufacturing industries.

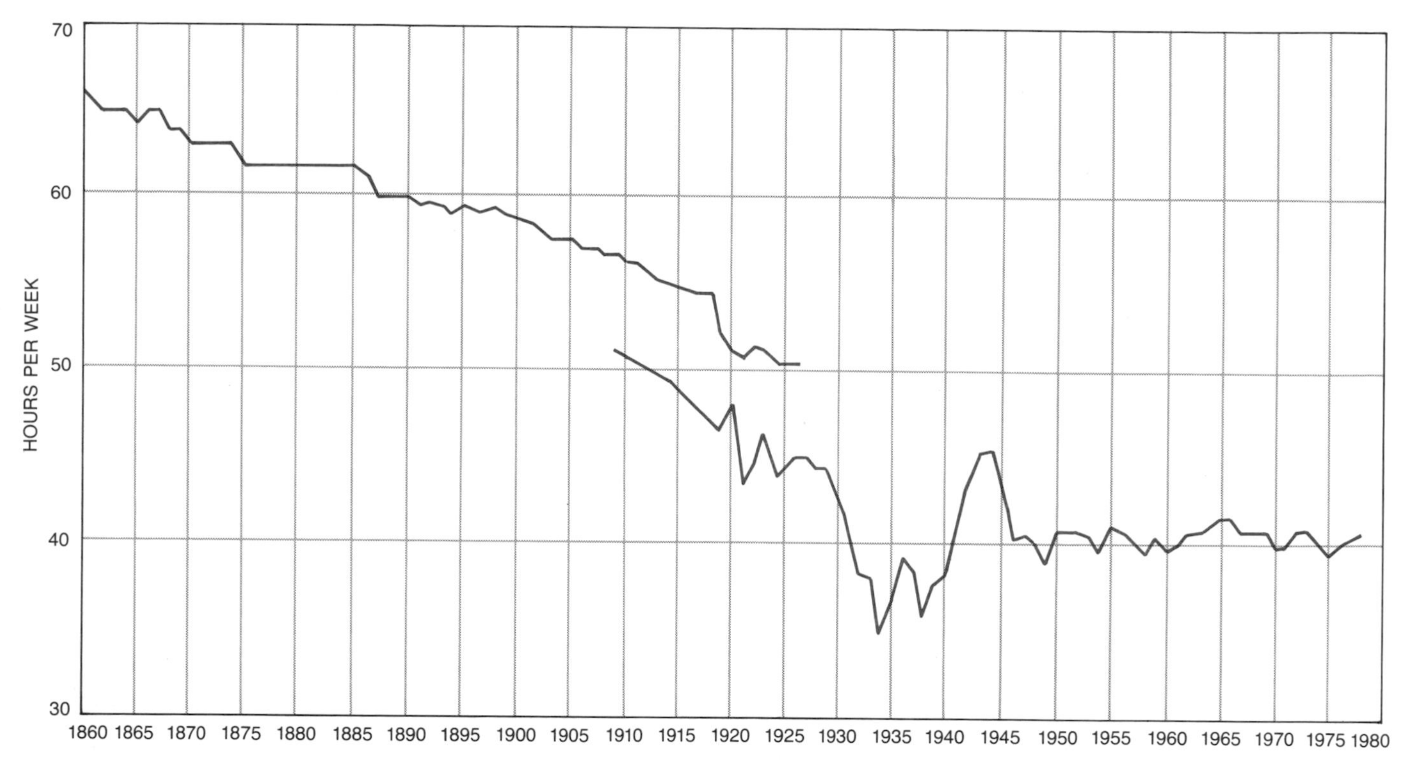

WORK WEEK IN MANUFACTURING INDUSTRIES of the U.S. shortened from about 67 hours in 1860 to about 42 hours in 1950 and has remained constant since then. Such reduction in the average number of working hours per week per employee amounts to the withdrawal from work of more than a third of the manufacturing labor force. The work week actually fell below 40 hours in the Great Depression of the 1930's with "sharing of unemployment" in part-time jobs and climbed well above 40 hours with overtime work in war production in the 1940's. The shortening of the work week together with income policies to maintain and increase, as the increase in output allows, the take-home income of the labor force constitutes one strategy for offsetting technological unemployment (*see illustrations on pages 107 and 108*). The discontinuity in the curve, over the period 1910 through 1925, reflects a change in the statistical time series kept by the country's bookkeepers involving principally changes in their accounting of the time of part-time and seasonal workers.

employs fewer than 50 men per shift.

Americans might have continued to absorb potential technological unemployment by voluntary shortening of the work week if real wages had risen over the past 40 years faster than they actually have, allowing the expectation of increase not only of total annual pay but also of total lifetime take-home pay. Because of the greatly expanded opportunities to replace labor by increasingly sophisticated technology it appears that the impersonal forces of the market no longer favor that possibility. Government policies directed at encouraging a steady rise in real wages sufficiently large to induce workers to resume continuous voluntary reduction in the work week could once have been considered. Under present conditions such policies would require such a large increase in the share of total national income going to wages that it would bring decline in productive investment, which is financed largely by undistributed corporate earnings and the savings of the upper income group. This would result in an unacceptable slowdown of economic growth. There remains the alternative of direct action to promote a progressive shortening of the work week combined with income policies designed to maintain and to increase, as increases in total output allow, the real family income of wage earners and salaried employees.

Recent studies sponsored by the U.S. Department of Labor seem to indicate that the total number of working hours offered by the existing labor force might be reduced in exchange for a more flexible scheduling of work time. Indeed, some workers, depending on their age group, family status, occupation and so on, would even be prepared to forgo a certain fraction of their current income, some by extension of their annual vacation, some by earlier retirement or sabbatical leave and some by working four and a half days per week instead of five. Reducing the work day by 15 minutes proves, incidentally, to be one of the less desirable alternatives. Tentative and obviously somewhat speculative computations based on the most desirable trade-off choices for different groups developed in these studies indicate that the average U.S. worker would be willing to forgo some 4.7 percent of earnings in exchange for free time. On the basis of the 1978 work year the average employee's work time would be reduced from 1,910 work hours to 1,821, or by more than two working weeks in a year.

Although such measures certainly deserve serious consideration and, if at all possible, practical implementation, they cannot provide a final answer to the long-run question of how to enable a modern industrial society to derive the benefits of continued technological progress without experiencing involuntary technological unemployment and resulting social disruption. Sooner or later, and quite probably sooner, the increasingly mechanized society must face another problem: the problem of income distribution.

Adam and Eve enjoyed, before they were expelled from Paradise, a high standard of living without working. After their expulsion they and their successors were condemned to eke out a miserable existence, working from dawn to dusk. The history of technological progress over the past 200 years is essentially the story of the human species working its way slowly and steadily back into Paradise. What would happen, however, if we suddenly found ourselves in it? With all goods and services provided without work, no one would be gainfully employed. Being unemployed means receiving no wages. As a result until appropriate new income policies were formulated to fit the changed technological conditions everyone would starve in Paradise.

The income policies I have in mind do not turn simply on an increase in the legally fixed minimum wage or in the hourly wage or other benefits negotiated by the usual collective bargaining between trade unions and employers. In the long run increases in the direct and indirect hourly labor costs would be bound to accelerate laborsaving mechanization. This, incidentally, is the explicitly stated explanation of the wage policies currently pursued by the benevolently authoritarian government of Singapore. It encourages a rapid rise in real wages in order to induce free domestic enterprise to upgrade the already remarkably efficient production facilities of this city-state. It is perhaps needless to add that these policies are accompanied by strict control of immigration and encouragement of birth control.

What I have in mind is a complex of social and economic measures to supplement by transfer from other income shares the income received by blue- and white-collar workers from the sale of their services on the labor market. A striking example of an income transfer of this kind attained automatically without government intervention is there to be studied in the long-run effects of the mechanization of agriculture on the mode of operation and the income of, say, a prosperous Iowa farm. Half a century ago the farmer and the members of his family worked from early morning until late at night assisted by a team of horses, possibly a tractor and a standard set of simple agricultural implements. Their income consisted of what essentially amounted to wages for a 75- or 80-hour work week, supplemented by a small profit on their modest investment.

Today the farm is fully mechanized and even has some sophisticated electronic equipment. The average work week is much shorter, and from time to time the family can take a real vacation. Their total wage income, if one computes it at the going hourly rate for a much smaller number of manual-labor hours, is probably not much higher than it was 50 years ago and may even be lower. Their standard of living, however, is certainly much higher: the shrinkage of their wage income is more than fully offset by the income earned on their massive capital investment in the rapidly changing technology of agriculture. The shift from the old income structure to the new one was smooth and practically painless. It involved no more than a simple bookkeeping transaction because now, as 50 years ago, both the wage income and the capital income are earned by the same family.

The effect of technological progress on manufacturing and other nonagricultural sectors of the economy is essentially the same as it is on agriculture. So also should be its repercussions with respect to the shortening of the work day and the allocation of income. Because of differences in the institutional setup, however, those repercussions cannot be expected to work through the system automatically. That must be brought about by carefully designed income policies. The accommodation of existing institutions to the demands and to the effects of laborsaving mechanization will not be easy. The setting aside of the Puritan "work ethic," to which Max Weber so convincingly ascribed the success of early industrial society, is bound to prove even more difficult and long drawn out. In popular and political discourse on employment, full employment and unemployment, with its emphasis on the provision of incomes rather than the production of goods, it can be seen that the revision of values has already begun.

The evolution of institutions is un-

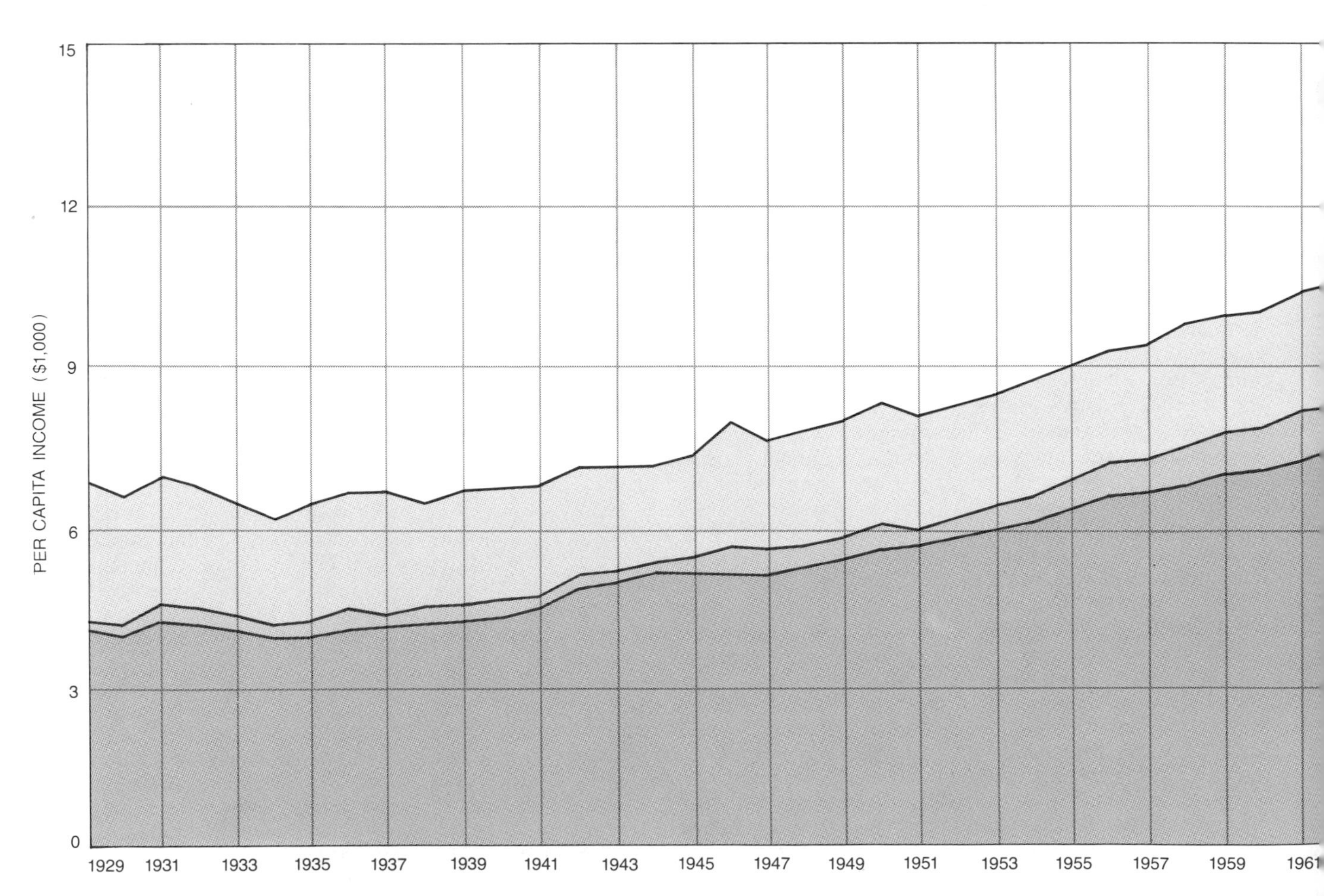

PERSONAL INCOME PER CAPITA in the U.S., plotted here in constant 1972 dollars, has more than doubled since 1929. The change in percentage shares of income accruing from property, transfer payments and labor (or to people receiving such income) reflects the evolution of the values and institutions of American society. The curves show that income from property has declined from about 40 percent to not much more than 15 percent of total personal income. Some of that decline reflects the exchange of profit and interest from small businesses (notably in trade and distribution and the services) for wages in large business enterprises (income from labor). It also reflects increased retention of earnings in corporations and increased financing of investment by such deflection of savings from person-

der way as well. In the structure of the tax system and through Social Security, medical insurance, unemployment benefits and welfare payments the country is finding its way toward necessary income policies. A desirable near-term step is to reduce the contrast between those who are fully employed and those who are out of work. This is the effect of the widespread European practice of paying supplemental benefits to those who work fewer than the normal number of hours per week. In the long run, responding to the incipient threat of technological unemployment, public policy should aim at securing equitable distribution of work and income, taking care not to obstruct technological progress even indirectly.

Implementation of such policy calls for close and systematic cooperation between management and labor carried on with government support. Large-scale financial transfers inevitably generate inflationary pressure. The inflation that dogs all the market economies, some more than others, does not arise from mere technical economic causes but is the symptom of deep-seated social problems. In this country it is basically the incessant wrangling between management and labor that keeps the cost-price spiral climbing.

West Germany, a country celebrated for its successful stabilization policies, is touted also as an example of the unregulated enterprise economy. In reality the success of the Schmidt government's anti-inflation measures rests on the firm foundation of institutionalized labor-capital cooperation in the management of German industry. The "codetermination" law requires that half of the board of directors of each large corporation be elected by labor, with the stockholders represented by the other half. Among the labor members some are "outside" directors representing the national trade unions. Since wage and employment questions constitute only one problem in the broad range of problems on the agenda of these boards, their deliberations bring employers and employees into working contact at the grass roots of German industry. That relationship cannot but be of crucial importance in determining the nature of agreements reached in collective bargaining conducted between the parties at the national level.

Austria is another country that has up to now successfully resisted inflationary pressure. Relations between management and labor are mediated by institutional arrangements very similar to those in Germany. The government plays a larger and more active role in the national across-the-board wage negotiations. It does so by contributing projections, drawn from the input-output data bank of the country's bookkeeping system, that link decisions affecting the industry in question to the situation of the country as a whole. This approach was employed, for example, to model and project the impact of the new text-processing and printing technologies on the Austrian newspaper industry. That technological revolution, the occasion for months-long disputes and work stoppages in Britain, the U.S. and other countries, was carried out smoothly and expeditiously in Austria by close cooperation between management and labor in accordance with detailed plans developed by the government. Until 1980, when the tidal wave of the second oil crisis, reinforced by the recession in the U.S. economy, reached Austria, the annual rate of inflation had been held below 4 percent and unemployment below 2 percent.

Although current business publications, trade papers and the popular press abound with articles about "automation" and "robotics" and speculation on the economic impact of these developments, only the governmental and scientific agencies of Austria have produced a systematic assessment of the prospective consequences of the present revolution in laborsaving technology in a modern industrial economy and society. That study, conducted for the government by the Austrian Academy of Sciences and the Austrian Institute for Economic Research, employed the country's input-output data bank to construct a model of the Austrian economy as of 1976. The model was then used to develop, in the words of Hertha Firnberg, the minister for science, in her introduction to the report of the study [see "Bibliography," page 116], "instead of unconditional prognostications—of either jubilation or horror—projections in the form of alternative scenarios... to analyze in quantitative terms the combined effects of economic, social and educational policy measures."

In input-output analysis the interindustry transactions that go into the production of the output of an economic system are arrayed in a matrix, with the outputs of each industrial sector displayed along its row and the inputs it draws from other industries in its column. The ratio of each input to the output of the sector—the input-output coefficient—reflects the technological requirement for that input which, although it is usually expressed in monetary value, is best visualized in the physical units appropriate to it, whether tons, bushels, barrels, kilowatts or man-hours [*see illustration on page 106*]. The entire column of input-output coefficients therefore presents the recipe of inputs required by the prevailing state of the technology involved in the production of that industry's product. At the foot of the column the human input is specified by the different kinds of labor supplied by the household sector.

For the Austrian study new sets of input-output coefficients had to be constructed reflecting changes in the input structure of all sectors of the economy prospectively dictated by the adoption of new laborsaving technology. In the simulation runs the effects of these changes could be gauged by comparison with the figures derived from actual interindustry transactions for 1976. Information for construction of the new coefficients was procured by comprehensive questionnaires circulated to technologists in each field and interviews with responsible technical directors of major industrial and service enterprises.

With all these data installed in the model, five alternative projections were run, describing in great detail the prospective state of the Austrian economy in the years 1985 and 1990. The sets of assumptions governing the projections differ from one another with respect to the rate of adoption of laborsaving technology, the extent of reliance on domestic as opposed to foreign suppliers of the new equipment, the more or less optimistic appraisal of the state of the world economy and last but not least the length of the work week for its effect on the distribution of employment among

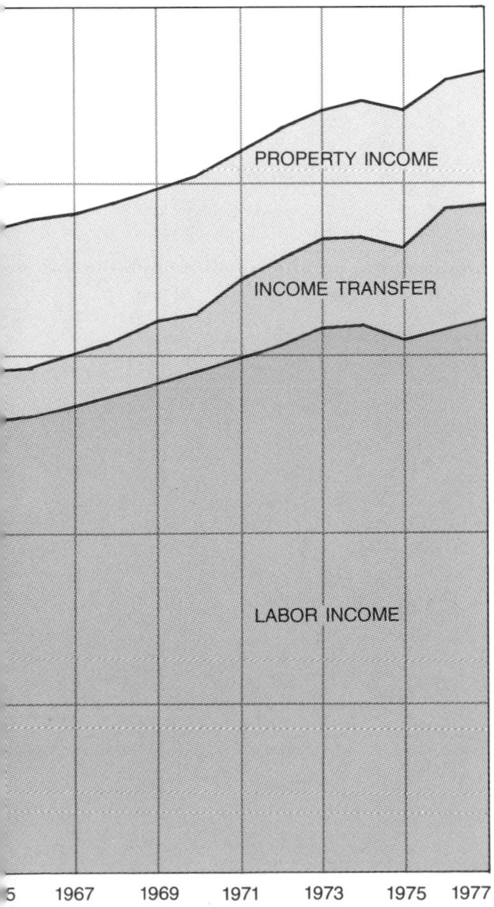

al income. Income from labor has increased from about 60 percent of the total to about 70 percent. Income from transfer payments (Social Security, medical benefits, unemployment compensation and so on) was negligible in 1929 but now is about 15 percent of total.

INPUT/OUTPUT COEFFICIENTS

		EXTRACTIVE	MANUFACTURING	HOUSEHOLDS	
EXTRACTIVE		.25	.40		
MANUFACTURING		.14	.12		
HOUSEHOLDS	LABOR	.70	2.80		
	CAPITAL	.10	.80		

		EXTRACTIVE	MANUFACTURING	HOUSEHOLDS	
EXTRACTIVE		.25	.40		
MANUFACTURING		.30	.20		
HOUSEHOLDS	LABOR	.40	1.00		
	CAPITAL	.81	1.25		

INPUTS AND OUTPUTS IN PHYSICAL UNITS

		EXTRACTIVE	MANUFACTURING	HOUSEHOLDS	TOTAL
EXTRACTIVE (BUSHELS)		25	20	55	100
MANUFACTURING (TONS)		14	6	30	50
HOUSEHOLDS	LABOR (HOURS)	70	140		210
	CAPITAL (TONS)	10	40		50

		EXTRACTIVE	MANUFACTURING	HOUSEHOLDS	TOTAL
EXTRACTIVE (BUSHELS)		31	32	62	125
MANUFACTURING (TONS)		38	16	26	80
HOUSEHOLDS	LABOR (HOURS)	50	80		130
	CAPITAL (TONS)	135	100		235

INPUTS AND OUTPUTS IN DOLLAR VALUES

		EXTRACTIVE	MANUFACTURING	HOUSEHOLDS	TOTAL
EXTRACTIVE		50	40	110	200
MANUFACTURING		70	30	150	250
HOUSEHOLDS	WAGES	70	140		210
	PROFITS	10	40		50
TOTAL		200	250	260	

		EXTRACTIVE	MANUFACTURING	HOUSEHOLDS	TOTAL
EXTRACTIVE		94	96	185	375
MANUFACTURING		150	64	106	320
HOUSEHOLDS	WAGES	50	80		130
	PROFITS	81	80		161
TOTAL		375	320	291	

INPUT-OUTPUT STRUCTURE of a rudimentary model economy is employed here to demonstrate the application of input-output analysis to assessment of the impact of mechanization on employment in an economic system. The two sets of three input-output tables show the system before mechanization (*left*) and afterward (*right*). For simplicity the model economy is disaggregated into two producing sectors, "Extractive" and "Manufacturing," and a "Households" sector that supplies labor and capital to the producing sectors; a real economy would be disaggregated into as many sectors as the data allow. The "Input-output coefficient" tables at the top of each set display the ratios of the inputs entered in the column for any sector in the tables of "Inputs and outputs in physical units," second from the top, to the total output entered at the end of the row for that sector in those tables. Thus the input-output coefficient table before mechanization, at the left, shows that the production of each ton of output from the manufacturing sector requires the input of .40 bushel from the extractive sector and .12 ton of its own product plus 2.80 hours of labor and .80 ton of capital stock (consisting of manufactured goods) provided by the households sector. Such coefficients may be derived from the record of actual transactions or from engineering data and other data and may be used to generate a new commodity flow table satisfying a different set of input demands in the households column. With prices at $2 per bushel, $5 per ton and $1 per hour and a return on capital at 20 percent of the value of the capital stock (physical units times price), the columns and rows in the "Inputs and outputs in dollar value" table, at the bottom left, can now be added and shown to balance: the value of the inputs equals the value of the outputs. The households row and column can be considered as being outside the interindustry matrix: the entries in its row correspond to the value-added of each industry; the entires in its column correspond to deliveries of each industrial sector to final demand. The equal totals of the households row and column correspond to the gross national product on the production and consumption sides respectively: $260 in the table at the left. In the tables at the right aggressive investment in laborsaving mechanization is assumed. Substantial increase in the coefficients for capital stock in both the extractive and the manufacturing sectors is reflected in increase in total capital stock from 50 tons to 235 tons in the physical units of the middle table. The consequent reduction in labor coefficients is reflected in the reduction of the labor input from 210 hours to 130. In spite of the increase in the price of extractive outputs to $3 per bushel, owing to increasing resource scarcities, the more efficient model economy now produces manufacturing outputs at the reduced price of $4 per ton and raises its gross national product to $291. With reduced employment, however, labor income falls from $210 to $130. The maintenance of consumption would therefore require deflection of income within the households sector from return on capital to labor income and income transfers, as in the real U.S. economy (*see illustration on pages 104 and 105*).

	1976 (ACTUAL)	1990 (PROJECTIONS)				
		UNCHANGED WORK WEEK		SHORTENED WORK WEEK		
		NO MECHANIZATION	FULL MECHANIZATION	NO MECHANIZATION	PARTIAL MECHANIZATION	FULL MECHANIZATION
GROSS DOMESTIC PRODUCT (10^9 SCHILLINGS)	738	1,180	1,190	1,113	1,114	1,148
INVESTMENT	197	365	365	365	365	365
PRIVATE CONSUMPTION	416	654	675	596	607	619
PUBLIC CONSUMPTION	133	172	174	162	163	168
EXPORTS	255	619	624	584	585	603
IMPORTS	262	631	648	595	605	606
GROSS DOMESTIC PRODUCT PER EMPLOYED PERSON (10^3 SCHILLINGS)	229	366	390	326	340	341
PER CAPITA WAGES (10^3 SCHILLINGS)	**101**	**150**	**159**	**131**	**136**	**137**
AVERAGE WORK WEEK (HOURS)	**42.1**	**39.6**	**39.9**	**35.2**	**35.3**	**35.3**
UNEMPLOYMENT (1,000 PERSONS)	**55**	**220**	**386**	**29**	**165**	**76**
EMPLOYMENT (1,000 PERSONS)	**3,222**	**3,221**	**3,056**	**3,413**	**3,277**	**3,366**
MEN	1,936	1,883	1,802	2,004	1,934	1,989
WOMEN	1,287	1,338	1,254	1,409	1,343	1,376

IMPACT OF MECHANIZATION ON ECONOMY of Austria in 1990 was projected in constant 1976 Austrian schillings by computer runs of a numerical matrix model of the input-output structure (*see illustration on page 106*) of the economy. The computer runs explored different sets of assumptions incorporated in the model to produce these columns of figures comparing the resultant differences in essential features of the system. The column at the far left shows the actual state of the economy in 1976. Under "Unchanged work week" the first column shows the economy projected to 1990 on the assumption of no change in the degree of mechanization already attained in 1976; the second column shows the economy projected on the assumption that mechanization employing technology demonstrated at the time of the study (1980) and embodied in largely imported equipment will have the full impact on jobs displayed in the table on page 162. Under "Shortened work week" the three columns from left to right show projections on the assumption (1) of no change from the 1976 state of mechanization, (2) of partial mechanization, that is, the realization (again with largely imported equipment) of 50 percent of the job impact displayed in the table on page 162 and (3) of full mechanization with equipment largely produced within the economy. Inspection shows that full mechanization with the work week unchanged yields the largest gross national product in 1990 (although only 10 million schillings larger than the projection based on 1976 mechanization), but it also causes the largest increase in unemployment. Shortening of the work week reduces the unemployment level in 1990 to around 2 percent even with full mechanization.

	BLUE-COLLAR				WHITE-COLLAR			
	JOBS AFFECTED	REDUCTION IN LABOR INPUT	JOBS AFFECTED 1990	REDUCTION IN EMPLOYMENT	JOBS AFFECTED	REDUCTION IN LABOR INPUT	JOBS AFFECTED 1990	REDUCTION IN EMPLOYMENT
AGRICULTURE AND FORESTRY	—	—	—	—	.01	.50	.200	.001
MINING	.68	.50	.072	.025	.10	.50	.038	.019
PETROLEUM	.60	.50	.235	.059	.20	.50	.076	.008
GLASS	.60	.50	.069	.021	.12	.50	.059	.008
FOOD PROCESSING	.55	.50	.114	.031	.10	.50	.154	.008
TEXTILES	.85	.67	.390	.222	.10	.50	.208	.010
CLOTHING	.89	.67	.210	.125	.07	.50	.177	.004
CHEMICALS	.55	.67	.300	.111	.20	.50	.206	.021
BASIC METALS	.73	.50	.369	.135	.13	.50	.182	.012
MACHINERY	.70	.77	.480	.259	.13	.50	.219	.014
METAL PRODUCTS	.80	.67	.215	.115	.13	.50	.195	.013
ELECTRICAL INDUSTRY	.65	.67	.700	.305	.13	.50	.220	.014
TRANSPORTATION EQUIPMENT	.50	.67	.352	.118	.13	.50	.219	.014
FOREST PRODUCTS	.75	.67	.075	.038	.10	.50	.118	.006
WOODWORKING	.75	.67	.075	.038	.10	.50	.118	.006
PAPER MANUFACTURE	.85	.67	.400	.228	.12	.50	.464	.028
PAPER PRODUCTS	.85	.50	.429	.182	.12	.50	.374	.022
CONSTRUCTION	—	—	—	—	.07	.50	.200	.007
ELECTRIC, GAS, WATER UTILITIES	.23	.50	.235	.027	.22	.50	.200	.022
TRADE	.53	.80	.100	.042	.18	.50	.200	.018
INFORMATION INDUSTRY	.41	.67	.020	.005	.11	.50	.200	.022
BANKS AND INSURANCE	—	—	—	—	.70	.50	.400	.140
HOTELS AND RESTAURANTS	—	—	—	—	.02	.50	.200	.002
OTHER SERVICES	—	—	—	—	.12	.50	.200	.012
HOUSING	—	—	—	—	—	1.00	0	—
GOVERNMENT	—	—	—	—	.64	.50	.180	.058

IMPACT OF MECHANIZATION ON JOBS in Austria is projected, industry by industry, from estimates made by engineers and other experts for an input-output study of the effects of mechanization on the Austrian economy (*see illustration on page 107*). The first column, under both the blue-collar and the white-collar headings, shows the percentage of jobs potentially affected by technology demonstrated as of 1980 although not yet installed on the production line or in the office; the second column shows the percentage of reduction of labor input in those functions potentially affected by such new technology; the third column shows the estimated percentage of jobs that would be displaced by 1990 if there were full application of the technology, and the fourth column shows the prospective percentage reduction in employment in 1990 that is the resultant of the other three percentages. Note the large percentage of blue-collar jobs potentially affected compared with the almost invariably small number of white-collar jobs affected, and the larger (in most cases) prospective reduction of labor input in blue-collar production functions compared with the uniform 50 percent reduction in white-collar office functions expected to result from the application of essentially the same technology to clerical and stenographic jobs in all industries and services.

different sectors and the rate of unemployment.

Out of the wealth of thought-provoking indications for the future to be found by close inspection of the several projections, it suffices for the purposes of the present discussion to cite just a few. The projections that carry the present state-of-the-art laborsaving technology into full application everywhere in the Austrian economy by 1990 lead in all cases to the largest increase in gross domestic product—but also to the highest levels of unemployment, to unemployment of 10 percent, a level not experienced in Austria since the dark days of the 1930's. With curtailment in the length of the work week at the maximum degree of mechanization, the direction of both the positive and the negative changes remains the same but their absolute magnitudes are reduced. Unemployment in this case comes closer to the civilized Austrian experience of 2 percent.

No comparable study has yet been completed for the U.S. economy. Fiscal starvation of the Federal statistical agencies has them currently sorting out interindustry-transactions data for 1977, with publication scheduled for not sooner than 1984. The Austrian study presents the best model available for projection of conditions in the U.S. of 1990. The Austrian economy is a mere 3 percent the size of the U.S. economy, but it too is highly industrialized and diversified. With some stretch of the imagination the Austrian projection of a high degree of mechanization supported by rapid expansion of domestic manufacture of all kinds of electronic products can be interpreted as indicating the structural changes the U.S. economy is likely to undergo in the next 10 or 15 years.

The time span covered by these projections is short. Moreover, they reckon with the consequences of the application of the state of the art of mechanization only as of 1980 at the latest, a state soon to be made obsolete by rapid advance in all the relevant technologies. These figures nonetheless throw some light on the quantitative dimensions of the profound challenge that an advanced industrial society must now begin to face under the impact of the continuing Industrial Revolution. History, even recent history, shows that societies have responded to such challenge with revision of their economic institutions and values conducive to the efficient use of changing technology and to securing its advantages for popular well-being. History shows also societies that have failed to respond and have succumbed to economic stagnation and increasing social disorder.

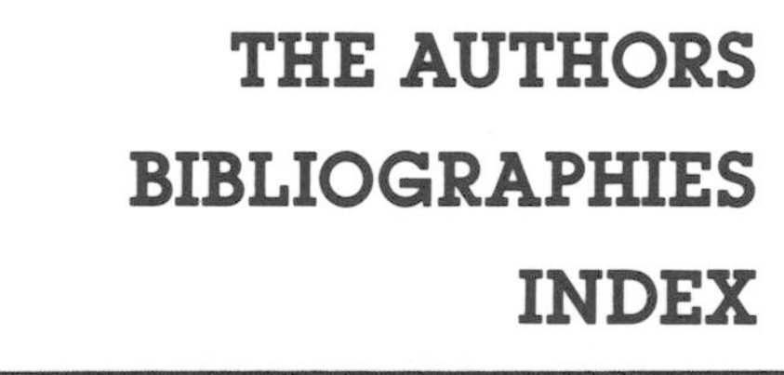
THE AUTHORS
BIBLIOGRAPHIES
INDEX

THE AUTHORS

ELI GINZBERG ("The Mechanization of Work") is director of the Conservation of Human Resources Project at Columbia University. He has been associated with Columbia since his undergraduate days. He earned his A.B. at the university in 1931, going on to obtain his A.M. in 1932 and his Ph.D. in economics in 1934. He joined the Columbia faculty in 1935. From 1967 to 1979 he was A. Barton Hepburn Professor of Economics at the Graduate School of Business. He retired in 1979, becoming emeritus professor and special lecturer. Ginzberg would like to express his gratitude to Anna B. Dutka for her help in the preparation of the current article.

WAYNE D. RASMUSSEN ("The Mechanization of Agriculture") is chief of the agricultural-history branch of the U.S. Department of Agriculture. His father was a Danish immigrant who ran a small cattle ranch in Montana, where Rasmussen grew up. In 1932 and 1933 he supported himself by teaching school while attending Eastern Montana College. He later transferred to the University of Montana, from which he received his B.A. in 1937. After being graduated he moved to Washington, where he joined the records-management division of the Department of Agriculture, remaining there until 1940. In the same period he was a student at George Washington University, from which he earned his M.A. in 1939. In 1940 he became a historian at the Department of Agriculture, and he has continued to work in the agricultural-history branch, with the exception of his period of service in World War II.

ROBERT L. MAROVELLI and JOHN M. KARHNAK ("The Mechanization of Mining") are respectively chief of the division of minerals availability and a staff engineer with the Bureau of Mines. Marovelli received his B.S. in mining engineering from the University of Alaska in 1950. After graduation he worked as an engineer for the Goodnews Bay Mining Company in Alaska. He then went to the Bureau of Mines, where his early work was on the abundance of iron, manganese and titanium in northern Minnesota; thereafter he also was concerned with blasting and rock structure. In the late 1960's he served as technology-research manager at the bureau's Twin Cities Mining Research Center, where he did mining research and special projects for other Government agencies (including work on lunar drilling for the National Aeronautics and Space Administration). In 1970 he moved to Washington as first chief of the Division of Mining Research—Health and Safety. Marovelli took up his present job this year. Karhnak's B.S. (1967) and M.S. (1970) in agricultural engineering are from Pennsylvania State University. From 1970 to 1976 he was a civilian employee of the Army, doing work on the hydraulics of heavy industrial and earth-moving machinery. In 1976 he moved to the Bureau of Mines, where he worked on improving the productivity of coal mining. In 1977 he was transferred to the Department of Energy, where he continued the work; in 1979 he returned to the Bureau of Mines.

THOMAS G. GUNN ("The Mechanization of Design and Manufacturing") is managing director of the computer-integrated manufacturing group at Arthur D. Little, Inc. He got into his current work by an unorthodox route. He was an undergraduate first at Antioch College in 1960, dropping out to drive racing cars. As an antidote to automobile racing, his father persuaded him to join the Army, and he served four years before returning to college at Northeastern University. After earning his B.S. in mechanical engineering in 1970 he went on to get his M.B.A. from Dartmouth College in 1977. He joined Arthur D. Little in 1979 following brief tours of duty in management positions in the shoe and computer industries. Gunn's nonprofessional interests include flying and house renovation.

MARTIN L. ERNST ("The Mechanization of Commerce") is vice-president, management science, of Arthur D. Little, Inc. He was graduated from the Massachusetts Institute of Technology with a B.S. in 1941. He then joined the armed forces; he worked for the Navy and Air Force until he joined the staff of Arthur D. Little in 1959. In 1941 he served as a physicist with the Naval Ordnance Laboratory and in 1942 he worked in the same capacity for the Naval Bureau of Ordnance. From 1943 until 1946 he was an operations analyst with the Air Force. In 1946 he joined the staff of the Cambridge Research Center, where he remained until 1948. He then moved to the office of the Chief of Naval Operations, ultimately becoming associate director. He left to join the operations-research section at Little. He is now head of the section in addition to holding his corporate position. He writes that he "became acquainted with the use of operations research during its very early military applications while working in Britain in 1942." Ernst adds that at Little "most of my work has been for the service industries, but for a wide variety of them rather than a single segment. My main interests have centered on the implications of the growing unification of computers and telecommunications."

VINCENT E. GIULIANO ("The Mechanization of Office Work") is a senior member of the information-systems section of Arthur D. Little, Inc. His bachelor's degree (1952) and his master's degree (1953) are from the University of Michigan. His doctorate (1959) is from Harvard University. In his graduate-student years Giuliano worked for the General Motors Engineering Development Laboratories, the Army Aberdeen Proving Ground and the Wayne State University Computation Laboratory. He joined the staff of Arthur D. Little in 1959 and has remained there ever since, with the exception of the period from 1967 to 1971, when he took a leave of absence to become the first dean of the Graduate School of Information and Library Studies at the State University of New York at Buffalo.

JOAN WALLACH SCOTT ("The Mechanization of Women's Work") is Nancy Duke Lewis University Professor and professor of history at Brown University. She was graduated from Brandeis University in 1962 and went on to obtain her Ph.D. in history from the University of Wisconsin at Madison in 1969. From 1970 to 1972 she was a member of the faculty of the University of Illinois at Chicago Circle; from 1972 to 1974 she was at Northwestern University, and from 1974 to 1980 she was at the University of North Carolina. In 1980 she moved to Brown. She is the coauthor with Louise A. Tilley of *Women, Work and Family* (Holt, Rinehart and Winston, 1978).

WASSILY W. LEONTIEF ("The Distribution of Work and Income") is professor of economics and director of the Institute for Economic Analysis at New York University. He was born in Russia in the city then called St. Petersburg. By the time he received his M.A. from the university there the city was known as Leningrad. He left the U.S.S.R. to study economics at the University of Kiel, from which he got his Ph.D. in 1928. He then worked at a variety of jobs in several parts of the world, including that of economic adviser to the Chinese government in Nanking. In 1932 he joined the faculty of Harvard University. He was at Harvard until 1975, ultimately occupying the Henry Lee Chair of Political Economy. At the end of his tenure at Harvard he went to New York University. His achievements in economics include the development of input-output methods of analyzing economies, a technique described in his work *Input-Output Economics* (Oxford University Press, 1966). For his work Leontief has received many awards and honors, including the Nobel prize in economics for 1973.

BIBLIOGRAPHIES

Readers interested in further explanation of the subjects covered by the articles in this issue may find the following lists of publications helpful.

THE MECHANIZATION OF WORK

GOOD JOBS, BAD JOBS, NO JOBS. Eli Ginzberg. Harvard University Press, 1979.

TECHNOLOGY AND SOCIAL CHANGE. Edited by Eli Ginzberg. Columbia University Press, 1979.

SERVICES/THE NEW ECONOMY. Thomas M. Stanback, Jr., and Thierry Noyelle. Allanheld, Osmun & Co. Publishers, Inc., 1981.

THE SERVICE SECTOR OF THE U.S. ECONOMY. Eli Ginzberg and George J. Vojta in *Scientific American,* Vol. 244, No. 3, pages 32–39; March, 1981.

THE MECHANIZATION OF AGRICULTURE

THE NEW REVOLUTION IN THE COTTON ECONOMY: MECHANIZATION AND ITS CONSEQUENCES. James H. Street. University of North Carolina Press, 1957.

POWER TO PRODUCE: YEARBOOK OF AGRICULTURE, 1960. U.S. Department of Agriculture, U.S. Government Printing Office, 1960.

THE AGRICULTURAL TRACTOR, 1855–1950. Roy B. Gray. American Society of Agricultural Engineers, 1974.

WHEREBY WE THRIVE: A HISTORY OF AMERICAN FARMING, 1607–1972. John T. Schlebecker. Iowa State University Press, 1974.

AGRICULTURE IN THE UNITED STATES: A DOCUMENTARY HISTORY. Wayne D. Rasmussen. Random House, Inc., 1975.

MECHANIZATION OF COTTON PRODUCTION SINCE WORLD WAR II. Gilbert C. Fite in *Agricultural History,* Vol. 54, pages 190–207; January, 1980.

THE MECHANIZATION OF MINING

MINERALS YEARBOOK. U.S. Department of the Interior, Bureau of Mines. U.S. Government Printing Office.

75 YEARS OF PROGRESS IN THE MINERAL INDUSTRY, 1871–1946. Edited by A. B. Parsons. American Institute of Mining and Metallurgical Engineers, 1947.

COAL—A CONTEMPORARY ENERGY STORY. Kristina Lindbergh and Barry Provorse. Scribe Publishing Corp. and *Coal Age,* 1977.

ECONOMIC MINABILITY OF HARD COAL OCCURRENCES IN THE WORLD—INTERRELATIONSHIPS AND DEVELOPMENTS. Gunter B. Fettweis in *Gluckauf,* Vol. 117, No. 16, pages 1019–1029; August 20, 1981.

THE MECHANIZATION OF DESIGN AND MANUFACTURING

ROBOTICS IN PRACTICE. Joseph F. Engleberger. Kogan Page and Avebury Publishing Company, 1980.

COMPUTER APPLICATIONS IN MANUFACTURING. Thomas G. Gunn. Industrial Press Inc., 1981.

IMPLEMENTING CIM. George H. Schaffer in *American Machinist,* Vol. 125, No. 8, pages 151–174; August, 1981.

THE MECHANIZATION OF COMMERCE

SWEEPING CHANGES IN DISTRIBUTION. James L. Heskett in *Harvard Business Review,* Vol. 51, No. 2, pages 123–132; March–April, 1973.

THE CONSEQUENCES OF ELECTRONIC FUNDS TRANSFER: A TECHNOLOGY ASSESSMENT OF MOVEMENT TOWARD A LESS CASH/LESS CHECK SOCIETY. Arthur D. Little, Inc. Report prepared for the National Science Foundation, Research Applied to National Needs (RANN), U.S. Government Printing Office, 1975.

THE RELATIONSHIP BETWEEN MARKET STRUCTURE AND THE INNOVATION PROCESS. Gordon Raisbeck and Mark Schupack. Prepared by Arthur D. Little, Inc., for the American Telephone and Telegraph Company, 1976.

THE WIRED SOCIETY. James Martin. Prentice-Hall, Inc., 1978.

NATIONAL TRANSPORTATION POLICIES THROUGH THE YEAR 2000. National Transportation Policy Study Commission. U.S. Government Printing Office, 1979.

THE MECHANIZATION OF OFFICE WORK

A HIDDEN PRODUCTIVITY FACTOR. Vincent E. Giuliano in *Telephony,* Vol. 199, No. 3, pages 30–36, 79; July 21, 1980.

ECONOMICS AND VALUES IN THE INFORMATION SOCIETY. V. E. Giuliano in *Information Society: Changes, Chances, Challenges.* 14th International TNO Conference. The Netherlands Organization for Applied Scientific Research TNO, 1981.

OFFICE AUTOMATION: A SURVEY OF TOOLS AND TECHNIQUES. David Barcomb. Digital Press, 1981.

THE MECHANIZATION OF WOMEN'S WORK

THE FEMALE LABOR FORCE IN THE UNITED STATES: FACTORS GOVERNING ITS GROWTH AND CHANGING COMPOSITION. Valerie Kincade Oppenheimer. University of California Press, 1970.

THE "INDUSTRIAL REVOLUTION" IN THE HOME: HOUSEHOLD TECHNOLOGY AND SOCIAL CHANGE IN THE 20TH CENTURY. Ruth Schwartz Cowan in *Technology and Culture,* Vol. 17, No. 1, pages 1–23; January, 1976.

WOMEN, WORK, AND FAMILY. Louise A. Tilly and Joan W. Scott. Holt, Rinehart and Winston, 1978.

WOMEN AT WORK: THE TRANSFORMATION OF WORK AND COMMUNITY IN LOWELL, MASSACHUSETTS, 1826–1860. Thomas Dublin. Columbia University Press, 1979.

THE ECONOMICS OF WOMEN AND WORK. Edited by Alice H. Amsden. Penguin Books, 1980.

THE DISTRIBUTION OF WORK AND INCOME

MACHINES AND MAN. Wassily Leontief in *Scientific American,* Vol. 187, No. 3, pages 150–160; September, 1952.

LIFETIME ALLOCATION OF WORK AND INCOME. Juanita Kreps. Duke University Press, 1971.

EXCHANGING EARNINGS FOR LEISURE: FINDINGS OF AN EXPLORATORY NATIONAL SURVEY ON WORK TIME PREFERENCES. Fred Best. Employment and Training Administration, U.S. Department of Labor, U.S. Government Printing Office, 1979.

MIKROELEKTRONIK: ANWENDUNGEN, VERBREITUNG UND AUSWIRKUNGEN AM BEISPIEL OSTERREICHS. Österreichisches Institut für Wirtschaftsforschung and Österreichische Akademie der Wissenschaften, with a foreword by Hertha Firnberg. Springer-Verlag, 1981.

INDEX